SEMICONDUCTOR WET PROCESS REVOLUTION

SEMICONDUCTOR WET PROCESS REVOLUTION

Device Application of Plating, CMP, Cleaning

Manabu Tsujimura

1985
COMEBOOK
2008
A COMEBOOK, INC., PUBLICATION

Published by COMEBOOK, Inc.

Semiconductor WET Process Revolution

By Manabu Tsujimura

ISBN : 978-89-957402-1-7

Published in 2008 by
COMEBOOK
134-17 Sungsan-dong Mapo-gu
Seoul 121250 Korea
Telephone(+82)2 337 5951
Fax(+82)2 375 6075
www.comebook.com

Preface

"Dr. Tsujimura, this is a good read. More than just a textbook for students. Why don't you publish it?" said Prof. Doi of Kyushu University in February 2007.

The book he referred to as a good read is a textbook for students based on my experience over the last 20 years or more. I have dedicated myself to the development of semiconductor manufacturing systems/components since I made my first step into the world of semiconductors in 1985. In 2002, I earned a doctorate under the supervision of Prof. Ota of Tokyo Metropolitan University. After that, I was unexpectedly invited as a visiting professor by Prof. Babu of Clarkson University, with whom I have conducted ongoing research on Chemical Mechanical Planarization (CMP) since 1997. Shortly afterward, Tokyo Metropolitan University and Prof. Park of Hanyang University both offered me visiting professors. I have given a series of lectures entitled "Wet Revolution" on semiconductor manufacturing systems in Japan, the U.S., and Korea. This textbook was born from these lectures.

For the last 32 years, I have been in charge of development and technical leader at Ebara Corporation. This book, inspired by my personal experience, is intended to be a handbook and provides general guidance for wet processes for semiconductor device manufacturing and their development, including polishing and plating. Target readers of this book are university students who study these processes/systems, as well as engineers who take charge of development for their companies. This book therefore contains both a tutorial part for students and an advanced part for engineers. Readers are advised to read the relevant part according to their requirements.

I believe that science and technology have evolved by 99 % imitation, 1 % innovation (something new). Mere imitation is regarded as an act of a lazy mind, but I think that imitation is a good start if we consider that the efforts of our predecessors have born the fruits of evolution, innovation, and revolution. After a period of imitation ends, we should take pride as an engineer in working hard to develop new technology. Though 99% of this book is based on the findings of our predecessors, I believe that it contains 1 % of something new. Comments on this book from readers are always welcome, except for harsh criticisms from engineers and students who have received bad grades from me.

I would like to express my special thanks to Dr. H. Shibata and Dr. M. Kodera of Toshiba Corporation for providing me with valuable data that were indispensable in the writing of this book. I would also like to thank Prof. Ota of Tokyo Metropolitan University, my supervisor for my doctorate earned in 2002, Prof. Babu of Clarkson University, the first person who invited me as a visiting professor, and Prof. Park of Hanyang University, the organizer of the Japan-Korea internship program, who acts as a bridge between Japan and Korea. I hope to receive their continued support in the training of students. My dream is to establish a fund (from my modest retirement allowance) for holding a conference entitled "The Evolution of the Wet Revolution" and to invite foreign students once a year after working as a visiting professor in five nations and retirement from Ebara Corporation. I hope that the professors help me to realize my dream (what nerve I have to ask such a favor...).

I attribute the readability of this book (I am one of those who think this book a good read) to my students. Various questions have been raised by students at Clarkson University. The seeds of development, as well as the seeds of this book, lay in such questions. Three to four internship students are invited from Hanyang University every year to be trained in Japan for 1 to 2 months; they have been so hard-working that I have needed to study more. Students at Tokyo Metropolitan University participated in experiments, including a role-play class entitled "Bell Best Project" described in this book. This book contains both what I taught to my students based on my service at Ebara Corporation and what I learned from my students.

Thanks to all who helped me write this book, including chipset/device manufacturers, material manufacturers, investment analysts, etc., as well as my colleague engineers who work hard around-the-clock.

Lastly, I must not forget to thank my beloved wife, Kumiko, who has been patient and helpful (she appears somewhere in this book).

July 2007
M. Tsujimura

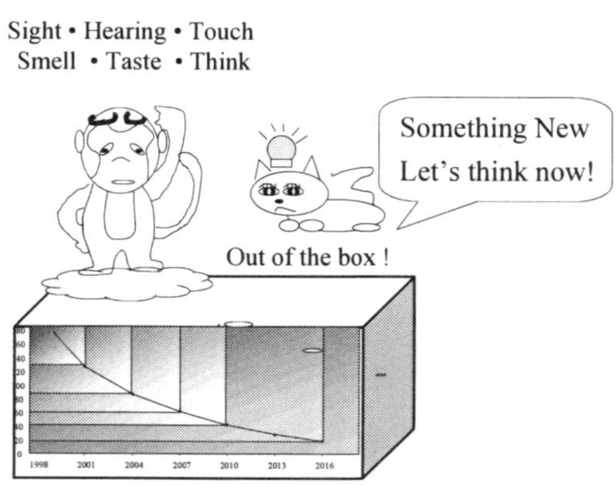

- Contents -

Chapter I

Planarization Carol

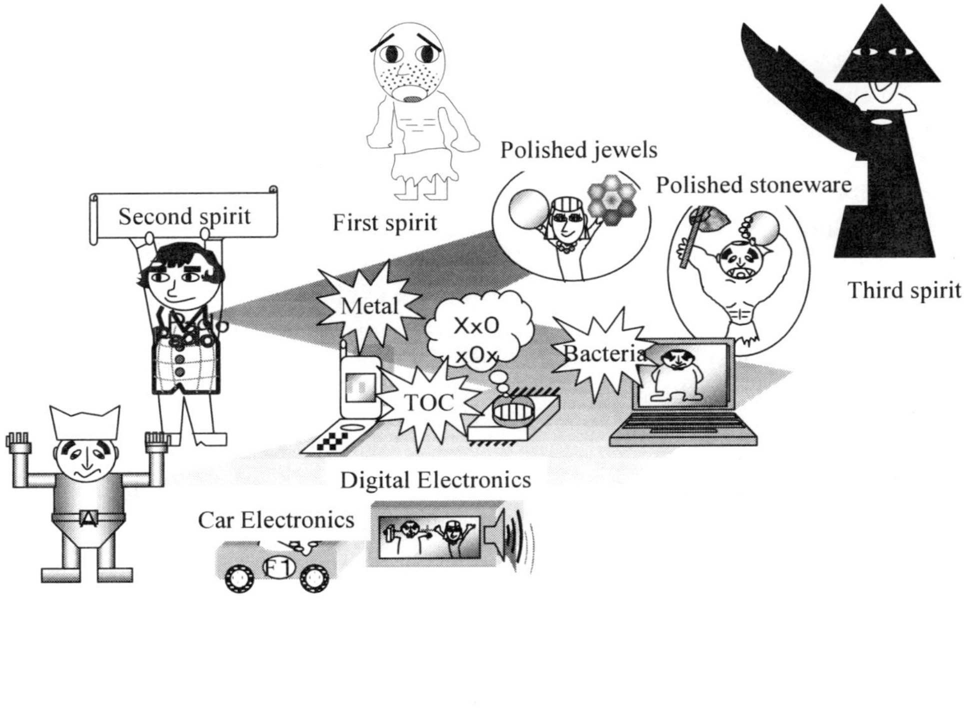

The oldest technologies in the world: Plating and polishing

If a man keeps cherishing his old knowledge, so as continually to be acquiring new, he may be a teacher of others (Analects of Confucius).

Nothing other than this quote would be suitable for describing plating and polishing technologies for manufacturing semiconductor devices.

Both polishing, as seen in stone tools, and plating, as seen in the Great Buddha in Nara, are indispensable technologies for manufacturing state-of-the-art semiconductor devices. In fact, their basic principles have not been changed since then, except in the size of objects to be worked – now they are *nano-sized*. This book introduces such old and new technologies.

Plating and polishing for semiconductors?

In the 1980s, IBM started using an astonishing process. It used copper (Cu), which had been avoided for semiconductor devices, and plating and polishing technologies, which were typically regarded as 3K technologies (kitanai (dirty), kitsui (tough), and keiken-izon (experience-dependent)) in Japan. Today, no state-of-the-art integrated circuit devices can be manufactured without plating and polishing technologies. Who could imagine the present situation in the past? I must confess that I did not trust either of them at all when I started using plating in 1987 and polishing in 1989. As I continued using them, I began to think that they had great potential. Why? You will find the reason later in this book. To put it simply, polishing offers excellent planarization performance. I still remember what one device engineer said: "Polishing is like a drug. Once you use it, you can never get away from it." After the introduction of polishing, plating began to be used for device manufacturing. Then, he said, "I might as well hang for a sheep (plating) as for a lamb (polishing)." Both comments sound dangerous; however, of course, the engineer meant nothing criminal.

Today's plating and polishing technologies owe a lot to bold semiconductor engineers like him, who are willing to use anything as long as excellent performance is provided.

Plating and polishing technologies supported by universities and academies

As explained above, old plating and polishing technologies began to be employed for state-of-the-art semiconductors. Oldness also means maturity. Although regarded as 3K technologies in Japan, they have been supported by fine old universities and companies. With this background, I would like to introduce how plating and polishing have evolved into advanced technologies.

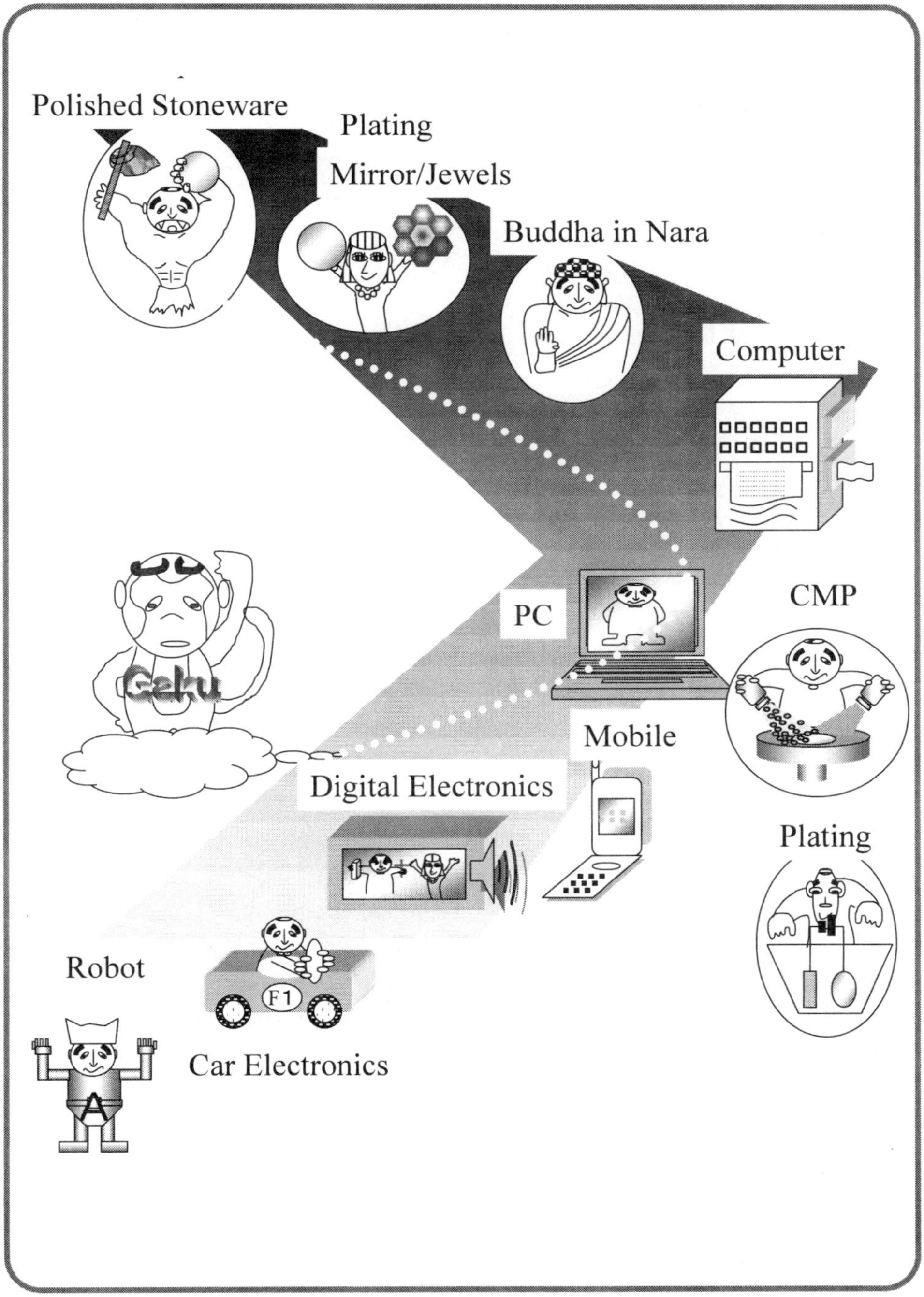
Polished Stoneware
Plating
Mirror/Jewels
Buddha in Nara
Computer
PC
CMP
Gaku
Mobile
Digital Electronics
Plating
Robot
F1
Car Electronics

Hello from Clarkson University

From Clarkson University

A 12-hour flight from Narita Airport to JFK International Airport. Another 90-minute flight to Burlington Airport. A 4-hour drive to Potsdam. Now, you arrive in the town where Clarkson University is located. Potsdam is a university town with a population of 13,000, most of whom are related to universities.

I have visited Clarkson to give lectures (or for research/conferences) for a week during the summer season once a year since I was invited as a visiting professor by the university for the first time in 2002. Why do I travel such a long way to visit such a small town (sorry!) every year? The answer is that Clarkson University is a center for CMP research.

With a 100-year history, Clarkson is a university that has made achievements in nanoparticle research over a long period of time. CMP research started as an application of nanoparticle research. Prof. S. V. Babu, former Vice Provost for Research and now honored with the title "Distinguished University Professor," has dedicated himself to the research. It would be no exaggeration to say that his former students have contributed to the introduction and growth of CMP technology at major device manufacturers, such as Intel, IBM, and Micron. It is said that you can find Clarkson alumni everywhere at any CMP conference.

Report from CAMP symposia in Lake Placid

Clarkson University organizes an annual international symposium on CMP. In 2007, the 12th International Symposium on Chemical-Mechanical Planarization was held in Lake Placid by the Center for Advanced Materials Processing (CAMP). For three days, engineers and scientists in the CMP field gathered from all over the world and discussed the latest topics of planarization technology for semiconductor devices.

This book is compiled of lectures I gave as a visiting professor at Clarkson University and presentations made for CAMP symposia. Therefore, the topics range from those suitable for students to those suitable for engineers of advanced devices. I also added an episode on dinner talks at CAMP (for families of symposium participants), which may entertain some readers.

That was a part of my experience at Clarkson University and CAMP. Before starting the main subject, I would like to introduce two fictional stories. Read them carefully since they contain the essence of plating, polishing, and the semiconductor industry. The details of the technologies are discussed later.

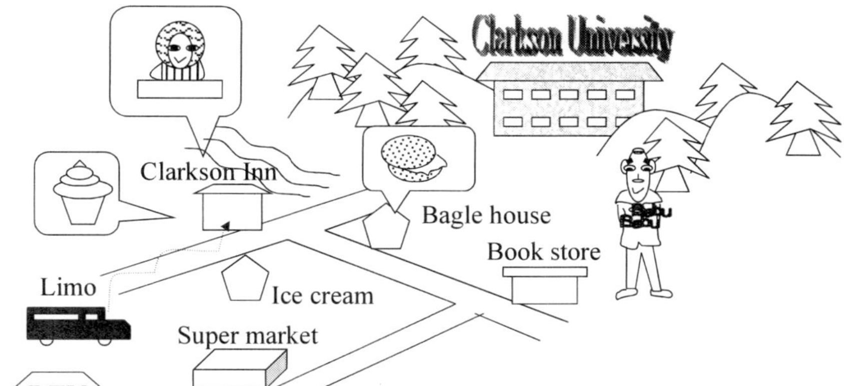

20 hours to Clarkson University

CAMP meeting

Planarization Carol

The first story is "A Planarization Carol." Never mind if you find that the story resembles Dickens's "A Christmas Carol," as I am a great fan of his story.

CMP booming!?

As I mentioned earlier, no other proverb but "If a man keeps cherishing his old knowledge..." suggests the basis of development. A combination of plating technology, known since B.C. 700, and polishing technology, which goes back to the Stone Age, brought a revolution in the semiconductor manufacturing technology in the 21st century. This is called a damascene process. CMP was developed as a planarization technology for inter layer dielectrics (ILD) by IBM in the early 1980s. IBM used the term *CMP* (chemical mechanical polisher) instead of simply *polisher*. How many of us have realized that naming is very important for development? If we simply call it a polisher, it still carries an old image of polishing. On the other hand, calling it CMP magically implies something new that can be used in new fields, such as semiconductors. CMP provides us with a good example of the development of something new through the study of our predecessors' assets.

The global CMP market in 2006 is estimated at US$ 1.4 billion (about 160 billion yen). The fields of application for CMP will continue to expand, and the market growth is promising. Will CMP players continue to enjoy their heyday? For your reference, the latest use of CMP for logic devices and memory devices is respectively shown, including ILD, tungsten contact, tungsten via, shallow trench isolation (STI), Cu interconnect, etc.

Is the future of CMP so bright? The semiconductor manufacturing industry has a history of rising and falling. Anything can happen at any time.

Attending a MRS meeting

I would like to introduce a wonderful experience I had. I was invited to a CMP symposium held by the Materials Research Society (MRS) in April 2007 and flew to San Francisco. This is a wonderful story of what I experienced at the symposium. Do you believe my story?

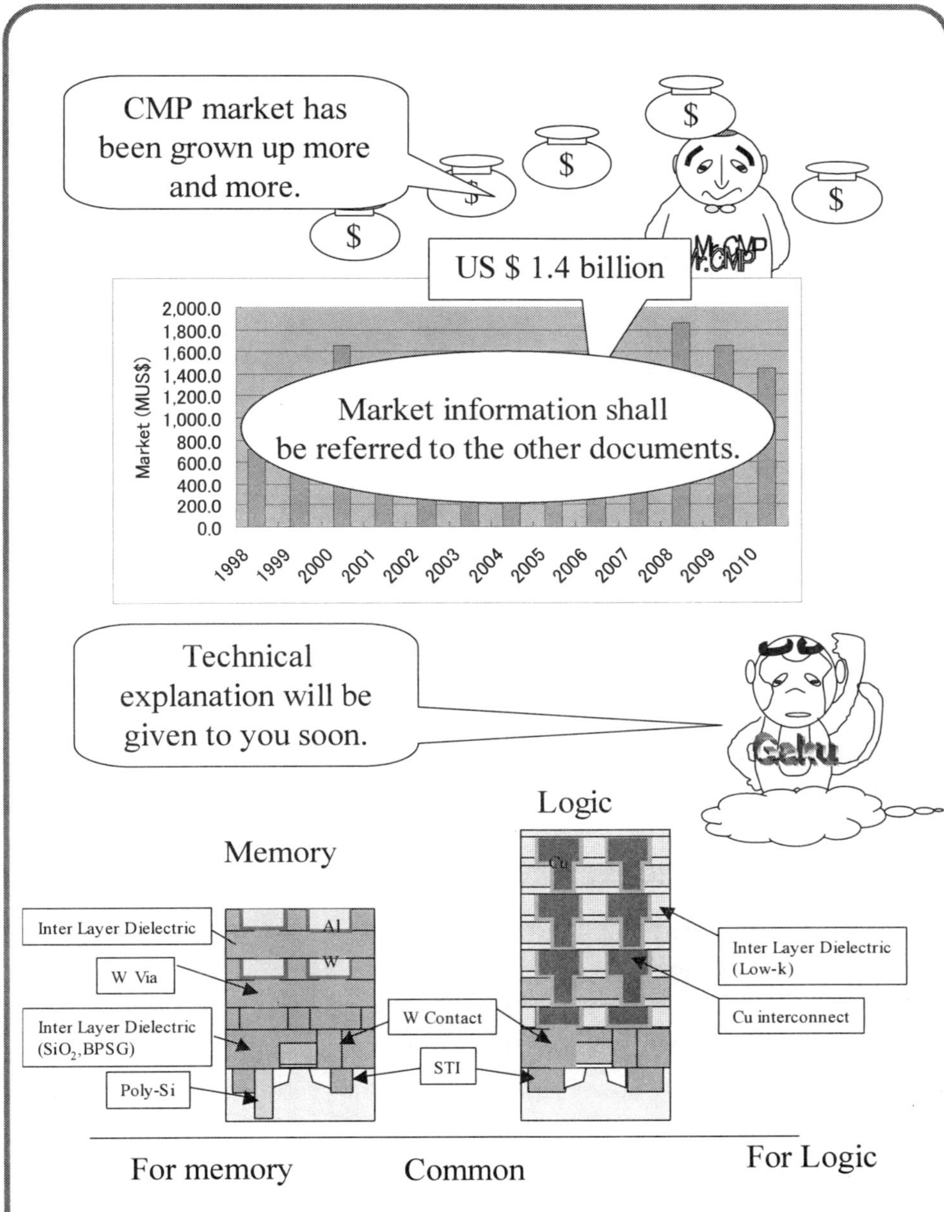

CMP application

Spirits appear

Three spirits

I didn't visit MRS for a long time. I have given a number of invited lectures since Prof. Babu of Clarkson University and I co-chaired a CMP symposium in 1998, but no CMP symposium has been held for the last few years. The CMP symposium in 2007 was held in full flourish for three days, attracting many attendants. However, fatigue began to make me fall asleep even during the symposium.

I often find that I cannot concentrate on the presentations in the symposium, thinking "another presentation by a student..." On that day, I thought I was listening, but as I drowsed...

An aged gentleman appeared

Suddenly, I found myself in darkness.

An aged gentleman with a familiar face came to talk to me.

The gentleman said, "Manabu, you look so happy with CMP now. For three days starting from tomorrow, spirits will come to you at eleven in the evening. Never doubt what they say and think it over."

I thought that I knew this gentleman from somewhere, but the fatigue resulting from jet lag made me fall asleep that day.

On the following days...

Note: Never mind if you have heard a similar story before.

I am a great fan of Dickens, and reading "A Christmas Carol" during my flight might have affected me.

First spirit (The Ghost of Polishing Past)

At eleven on the first day, the first spirit appeared as predicted. He looked like a primitive man. As soon as he said, "Mana~bu, I will take you to the past," the savage spirit grabbed me by my hair and took me to the Stone Age in a blink. There, I saw our ancestors polishing stone with stone. Men were working hard to shape stone tools, and women were making stone beads by polishing stone.

The spirit said, "Mana~bu, polishing is a technology based on the hard work of your ancestors and handed down since the beginning of time, none of which you have developed yourself. Don't forget your ancestors' technology!"

Students presentation?
What is CMP?
Manabu
Oh! It becomes dark suddenly.
Looks like an old gentleman.
Manabu
First spirit comes
To the past world
Polished stone tools
Polish stone beads

Past, present, and future

Second spirit (The Ghost of Polishing Present)

On the second day, another spirit looking like a modern man appeared.

As soon as the spirit said, "Manabu, how was the *Ghost of Polishing Past* you saw yesterday? I am the *Ghost of Polishing Present*. Let's see how CMP is used now," I found myself in darkness. "Where am I now?"

I looked around and found bacteria and germs all over. People wearing dirty clothes labeled as every organic matter and metal were running about.

The spirit said, "This is the inside of the *stone (device)* polished with *a* modern polisher. Don't be so arrogant as to think that CMP has already been perfected. CMP still has many problems, such as corrosion, scratches, delamination, and voids. The *talking stones* of the modern age are crying out, "Help us!""

Third spirit (The Ghost of Polishing Yet To Come)

On the third day, the last spirit appeared. He was clad in a black cloak and stood in silence.

I said, "Mr. Spirit, what are you going to show me? On the first day, I met our ancestors who worked hard for polishing since the beginning of time. Yesterday, I met the *talking stones* that were tortured on the cutting edge of semiconductors. What can I see today?"

The spirit in a cloak pointed in silence. He pointed diligent young engineers and scientists sweating on the front lines trying to explain state-of-the-art technology at a CMP symposium.

The spirit said, "Manabu, the present prosperity of CMP is a result of our predecessors' hard work and the efforts of young engineers in polishing technology in each period. In Japan, the Planarization Committee has devoted itself to promoting CMP since 1994. All of such people are trying to polish and clarify their *intent*. Starting with ILD, followed by tungsten, poly-Si films, STI, Cu damascene, and new processes, such as metal damascene gate, CMP applications have been steadily expanding. Where there's science, there's development. All of you involved in CMP technology, never be arrogant in the present prosperity, but polish your intent and make further efforts."

As soon as he finished his statement, he took off the cloak. It was the gentleman who appeared when I was drowsing.

I said, "Oh, you are..."

Yes, he was Prof. Babu, as anyone involved in CMP technology knows.

Second spirit comes

Third spirit comes

CMP forever

Together with Prof. Babu

Prof. Babu woke me up.

The semiconductor industry has a history of rising and falling.

CMP may not enjoy the same prosperity in the future.

In fact, the paradigm is significantly shifting around us [1)].

Both CMP and plating technologies have shifted the former paradigm. They must have beaten and superseded many other technologies behind this prosperity. Plating and CMP are no exceptions.

Note 1) *Paradigm Shift 45* is my alarm over a major paradigm shift brought about by scaling of 45 nm and beyond, thinner films of 45 μm or less, and larger wafer diameters of 45 cm or more, which began in 2005. Details will be described later.

CMP forever

The above were the contents of the dream I had on the occasion of my business trip. Putting the true meaning of this dream aside, nobody can stop the rising tide of CMP now. However, as I am repeatedly saying, every revolutionary technology has a history of rising and falling. All of us involved in CMP technology should avoid arrogance in today's prosperity, and continuously strive to promote further advancement of CMP. Once the "P" of CMP stood for polisher. In the MRS meeting in 1998, Prof. Babu and I decided to use the "P" for planarization. Polishing is not the only planarization technology.

There are so many issues to be addressed in planarization technology. The technology is supported by young engineers, like those attending CAMP symposia. Now I am wide-awake. On my honor, I shall not be arrogant in today's prosperity, but do my best to see that CMP enjoy its prosperity 10 and 20 years from now.

CMP forever!

This is the first fictional story telling you the history and present state of plating and CMP. It has been favorably received by my students. How did you find it?

Next, I would like to introduce the second fictional story, which contains more advanced topics, so that you can get an overview of the present state of the semiconductor industry.

Now I am wide-awake.
What is CMP?
Manabu
Bah~!
Oh! You are Babu
Manabu Do you understand?
Paradigm is shifted.
P- 45 nm No More Moore!
It cannot be measured by light wave length.
Resistivity of Cu may be bigger than Al?
Wire
ρ
P-45 μm No More Thinner
Difficult to make thinner that 45 um.
SOB
SiP
SOC
P-45 cm No More Lar
$
Who shares the R&D cost?
Non-uniformity may be bigger.
450 mm
300 mm
Investor
Engineer
Scientist

Project X?

I am trying to explain, "What is a chipset? What is a device? What is a system? What is the semiconductor industry?" by introducing a robot development project implemented by a fictive chipset manufacturer, device manufacturer, and tool manufacturer. The details of individual technologies will be described in later chapters. The purpose of this section is to glimpse the overall picture.

Bell project takes off!

This story features the following characters.

- Dr. Manabu, CEO of Sunny Corporation, a robot development company (a very stubborn and headstrong person)
- Ms. Kumiko, Manager of TBS Corporation, a device manufacturing company (a very capable and good-looking person)
- Mr. Babu, Manager of BER Corporation, a tool manufacturing company (a very understanding engineer)

The story begins at an office of Sunny Corporation on one afternoon.

A robot development project is started by Dr. Manabu, CEO of Sunny Corporation, who wishes to produce a robot like Astro Boy. What is going to happen?

The story consists of eight scenes as follows:

- Scene 1: I want Astro Boy!
- Scene 2: Food chain
- Scene 3: Bell Best Project
- Scene 4: Concept first
- Scene 5: Roadmap
- Scene 6: Bell device development (plating and polishing)
- Scene 7: Order!
- Scene 8: Bell Boy!

Let's start the story in the hope that you read it to the end without getting bored.

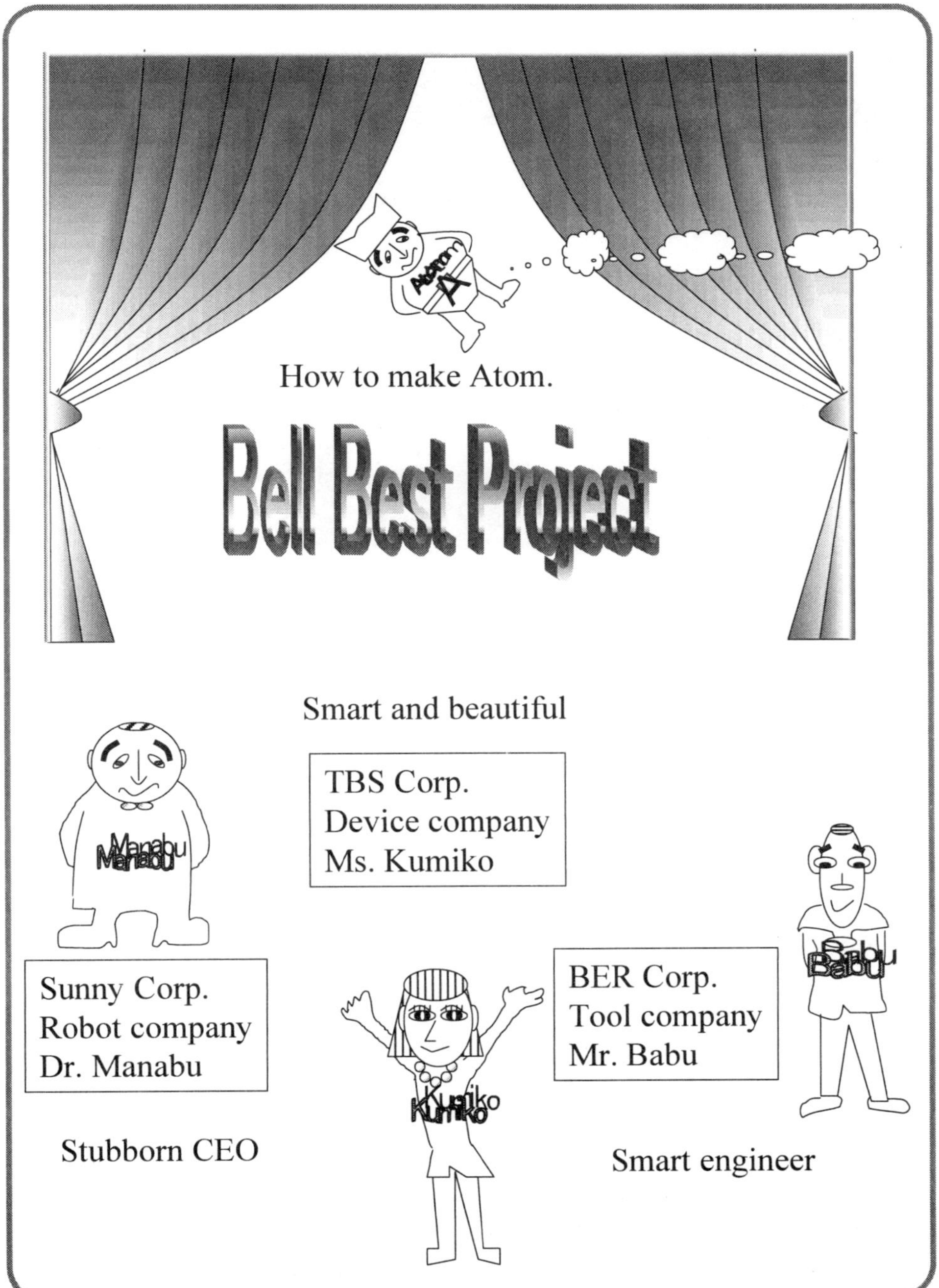
How to make Atom.
Bell Best Project
Smart and beautiful
TBS Corp.
Device company
Ms. Kumiko
Manabu
Babu
Sunny Corp.
Robot company
Dr. Manabu
BER Corp.
Tool company
Mr. Babu
Kumiko
Stubborn CEO
Smart engineer

Now is the time for Astro Boy

Scene 1: I want Astro Boy!

In the CEO's office at Sunny Corporation, Dr. Manabu, the stubborn CEO of the company, is always thinking about the future. Today, he is reading "Astro Boy."

Dr. Manabu: Technological innovation shifted from cars then to semiconductors, and now to biotechnology. What comes next? Yes, robots! I want an almost-human robot like Astro Boy, not machine-like robots like ASIMO and Gigantor. Development of semiconductor devices is indispensable to the development of such a robot. Well, I'll set out a development project for the robot of the future that will be as intelligent as I am!

Where is the project headed?
Here is some supplementary explanation of post-semiconductor technology.

What comes next after semiconductor technology?

The semiconductor industry has grown by adapting technologies from other mature industries. Now, it is the semiconductor industry's turn to make its technology available to new fields. But which fields?

Leading candidates are micro-electro-mechanical systems (MEMS), nanotechnology, and biotechnology. Japan's science and technology policy can be measured by budgets announced by the Ministry of Economy, Trade and Industry. In addition to energy and the environment, which are given priority in the national policy, budgets for information technology, biotechnology, and nanotechnology have been substantially increased for the last few years. For long-term research and development, we should make efforts to adapt information technologies developed for the semiconductor industry to biotechnology and nanotechnology.

What are the technologies developed for the semiconductor industry then? How various technologies, including microfabrication technology, nanofabrication technology, general system technology, and process technology, should be adapted to new fields is important. Digital home appliances in the short term, on-vehicle equipment in the middle term, and biotechnology, nanotechnology, robots, and their applications in the long term will each brighten the future of semiconductors. Semiconductors forever!

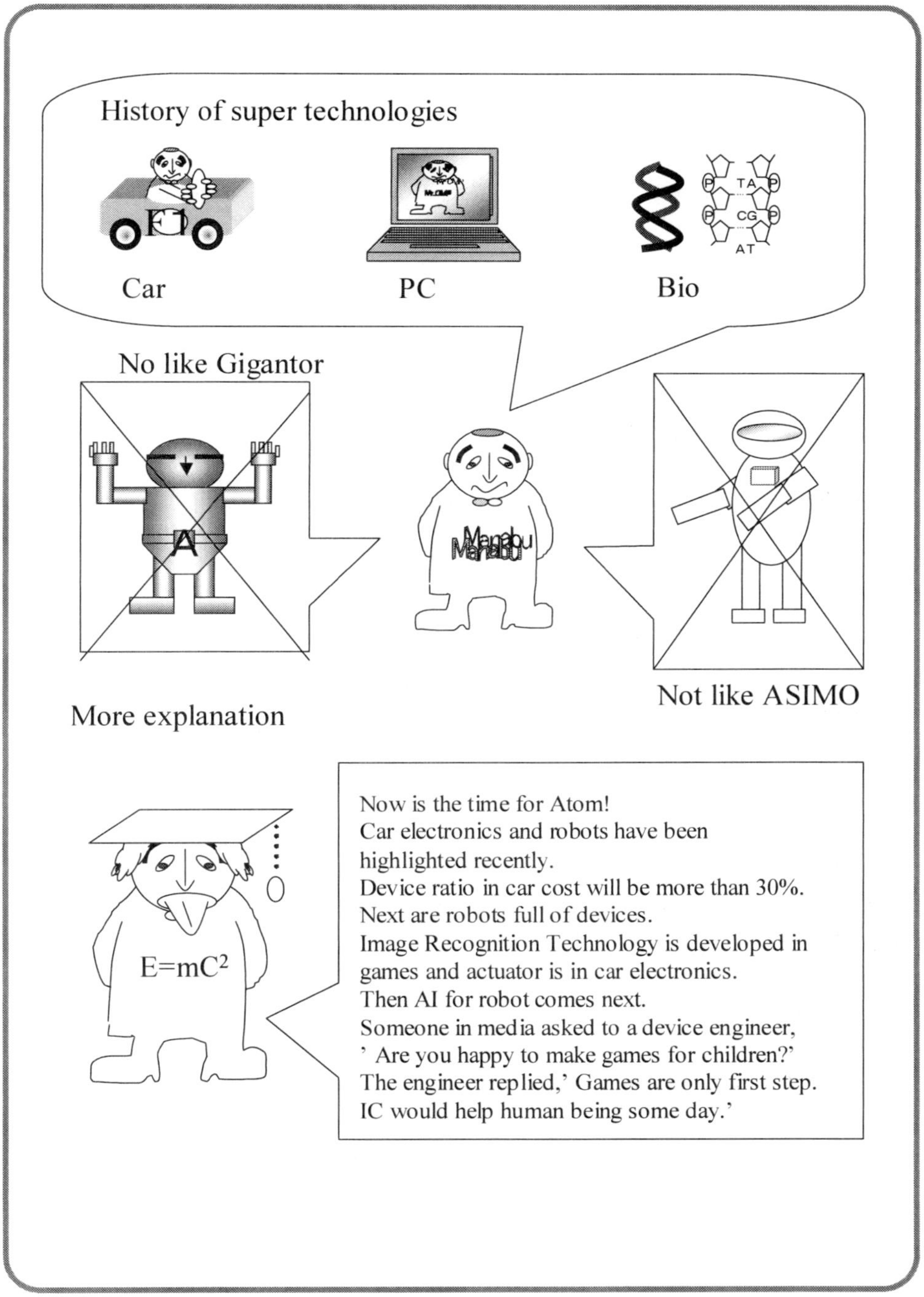
History of super technologies
Car
PC
Bio
No like Gigantor
Not like ASIMO
More explanation
$E=mC^2$
Now is the time for Atom!
Car electronics and robots have been highlighted recently.
Device ratio in car cost will be more than 30%.
Next are robots full of devices.
Image Recognition Technology is developed in games and actuator is in car electronics.
Then AI for robot comes next.
Someone in media asked to a device engineer, ' Are you happy to make games for children?'
The engineer replied,' Games are only first step. IC would help human being some day.'

Food chain in the semiconductor industry

Scene 2: Food chain

The story continues from Scene 1. Mr. Manabu calls his secretary.

Dr. Manabu: TBS is best suited for device development. Get Ms. Kumiko on the phone right away. She is promising to develop an excellent System on Chip (SoC). And, BER is suitable for process tool development. Make a call to Mr. Babu. I know he can meet all our needs.

The secretary is busy making arrangements. In any case, instructions are issued from the upstream side to the downstream side of the food chain in the semiconductor industry.

Note) Please note that Sunny, TBS, and BER are not real companies.

The food chain (supply chain) in the semiconductor industry consists of three sectors: chipset -> device -> manufacturing tool/material. I first used the term *food chain* at the International Trade Partners Conference (ITPC), a meeting of VIPs in the semiconductor device industry, to express the jungle of business. It is my own coined term, but seems to have been recognized in the industry.

Overview of the semiconductor industry

The market share of the device sector is about 20 % of the chipset market (or, semiconductor devices account for about 20 % of components in a chipset), while the share of the manufacturing system sector is only 2 % of the device market (accounting for 10 % of investment in the device sector). As the saying goes, "it is an ill wind that blows nobody any good." A change of 2 % in the chipset market results in a change of 20 % in the device market and further results in a change of 100 % in the manufacturing system market. This phenomenon is sometimes called the *bullwhip effect*.

Indeed, the semiconductor industry is dynamic and complex.

Food Chain

Set		Device		Tool
Sunny Corp. Robot company Dr. Manabu	⇨	TBS Corp. Device company Ms. Kumiko	⇨	BER Corp. Tool company Mr. Babu

Market Image: Upstream is influential with downstream

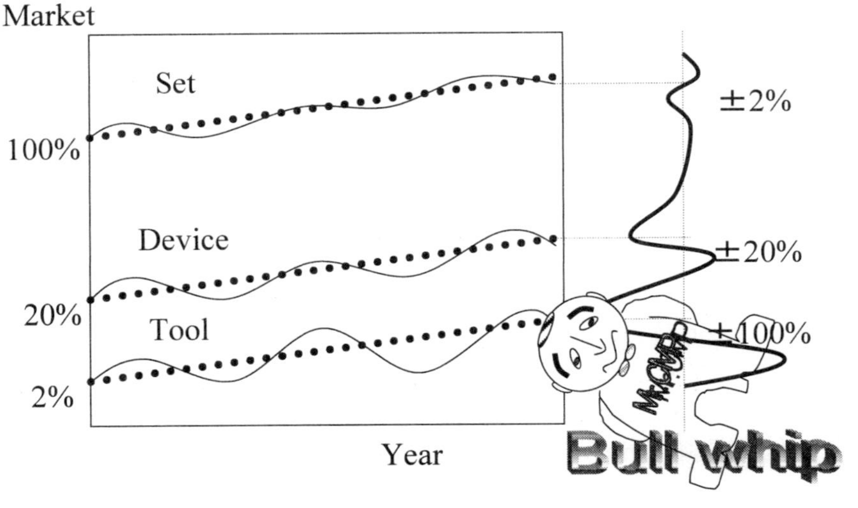

The project proceeds

Scene 3: Bell Best Project

By the instruction of Dr. Manabu, all project members gather for a meeting.

Dr. Manabu: The name of the device to be developed is important. I think it should be, for example, *Cell*.

Ms. Kumiko: Really? It's the name of the device developed by Toshiba and Sony!

Dr. Manabu: Oh, right! Then, how about *Bell*?

Ms. Kumiko: Ah, it sounds similar to..., but I think it's good.

Dr. Manabu: So, next, the project name! Propose any name.

Mr. Babu: Well, how about *Bell Best*?

Dr. Manabu: OK, OK, OK…

Thus, the Bell Best Project is launched.

In the actual field, projects are named in a manner as described above and proceed with insistence, some compromise, and obsequiousness.

The Bell Best Project has taken the first step toward robot development.

Scene 4: Concept first

Dr. Manabu wants to make a robot that resembles him and is sophisticated like Astro Boy. He sketches the mechanical design of the robot himself. A mechanical design is very important since it represents the primary design concept. Although, in some cases, mechanical designs are created after product design, mechanical designs created by designers should come first.

The figure on the next page shows the mechanical design created by Dr. Manabu. It cannot be said that his action is really based on the concept-first approach. In addition, he selects the suppliers himself. In fact, he is very dictatorial.

The brain, feet, body, and software of the robot are made by TBS, XX, TT, and ZZ, respectively. Then, the name of the robot is…

The name is Bell Boy after the device name Bell (what an easy name!).

The project proceeds according to the following outline.

- The name of the device product to be developed is Bell.
- The project name is Bell Best Project.
- The name of the robot to be made is Bell Boy.

So many discussions in vain?

Concept First ! This is my job!

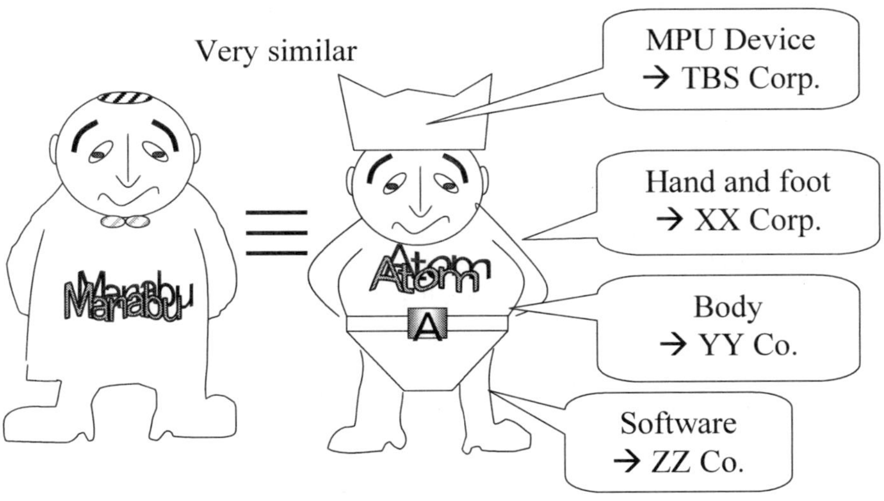

Creating roadmaps

Scene 5: Roadmap

Roadmaps are different from schedules. The key is to clarify critical project phases for individual project members, who in turn are expected to be committed to the completion of the phases.

Roadmap for the chipset manufacturer

The roadmap created by Sunny (chipset manufacturer) is the most important. Sunny performs market research and decides to release Bell Boy around Christmas 2010. To achieve this target, the Bell device must be developed by the end of 2009. This schedule is reported to TBS (device manufacturer). Dr. Manabu also decides to hold the kickoff meeting (first project meeting) on January 1, 2007 (how dictatorial he is!).

Roadmap for the device manufacturer

Based on the reported delivery schedule, TBS plans a detailed schedule for device development. Considering the period of design, processing, and manufacturing, TBS sets the end of 2008 as the delivery deadline for the manufacturing systems. After determining systems necessary for device manufacture and their development periods, it reports them to BER (tool manufacturer). To ensure that the deadline is met, TBS may set a slightly earlier period for the development of each tool.

Lithography tool to be developed by (date A)
Chemical vapor deposition (CVD) tool to be developed by (date B)
Reactive ion etching (RIE) tool to be developed by (date C)
CMP tool to be developed by (date D)
Plating tool to be developed by (date E)
Cleaning tool to be developed by (date F)
Exhaust tool to be developed by (date G), etc.

Roadmap for the tool manufacturer

Finally, BER sets its own schedule and makes its best effort to meet the deadline specified by TBS. Although common systems are easy to develop, development of unprecedented systems is difficult. A developed system is first verified internally (α version) and then verified on the device manufacturer site (β). Then, the final version (γ) is delivered.

Since cutting-edge functions are required for semiconductor manufacture, a β version is often delivered before γ version development.

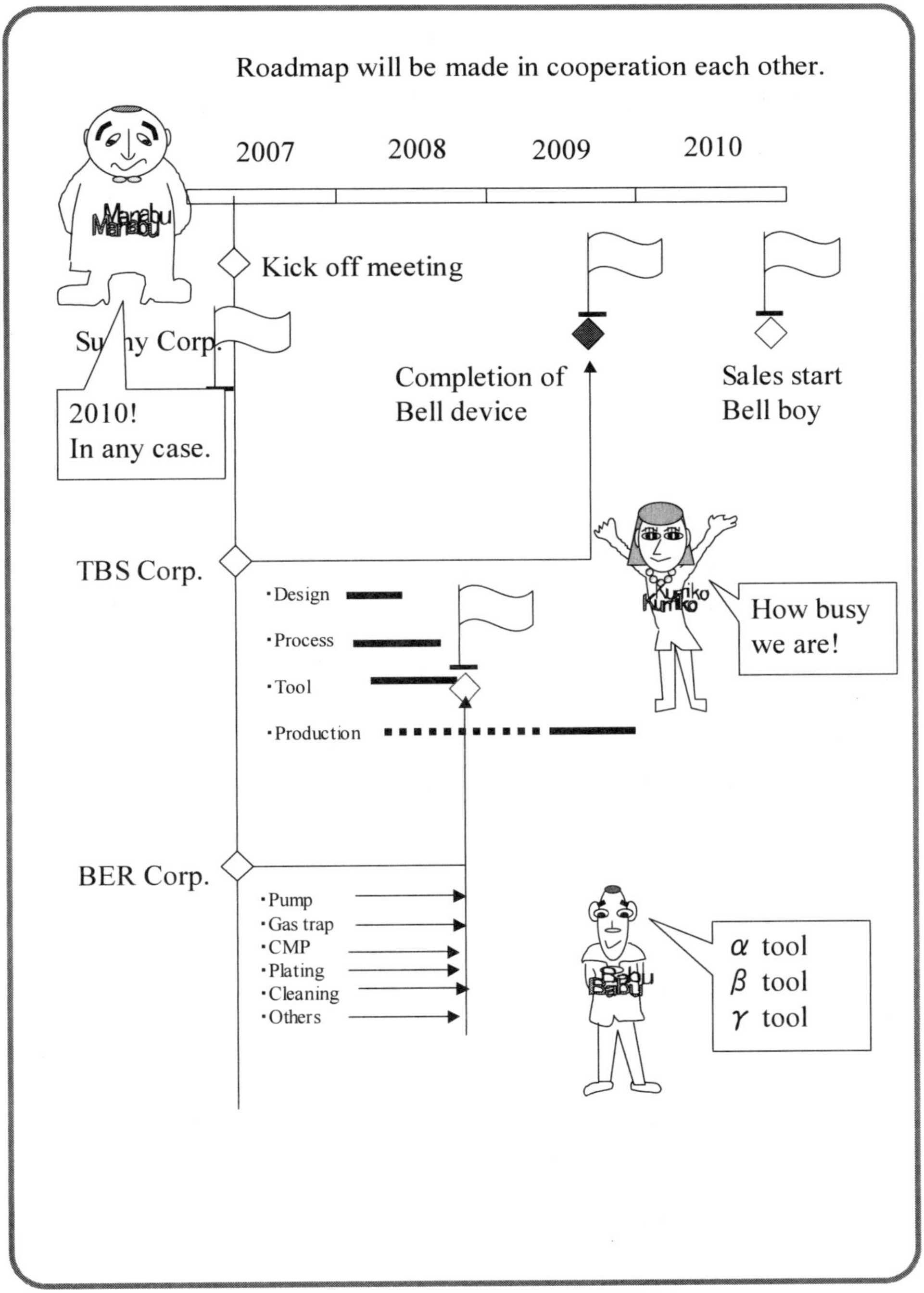
Roadmap will be made in cooperation each other.
2007
2008
2009
2010
Kick off meeting
Manabu
Su ny Corp.
2010!
In any case.
Completion of
Bell device
Sales start
Bell boy
TBS Corp.
·Design
·Process
·Tool
·Production
Kumiko
How busy
we are!
BER Corp.
·Pump
·Gas trap
·CMP
·Plating
·Cleaning
·Others
Babu
α tool
β tool
γ tool

What is TEG?

Scene 6: Bell device development (plating and polishing)

In a conversation between Mr. A, the CMP engineer, and Mr. Babu;

Mr. Babu: The demonstration results are not good. You've been instructed by the user about the initial profile and the substrate structure, haven't you?

A: What? What do you mean by *initial profile*? I don't have such a measurement item. And, the substrate structure is confidential (he sounds pretty sure of himself).

Mr. Babu: (Stunned by Mr. A's comment) But, but, can you guarantee polishing performance without knowing the initial step height or the pattern density?

A: …

In another conversation between Mr. B, the plating engineer, and Mr. Babu;

Mr. Babu: The demonstration results are not good. I suspect the TEG structure for plating specified by the user is too complicated. Or, is the pattern design so special?

B: What? What do you mean by *TEG for plating*? I've been informed of no pattern. It's confidential (he also sounds pretty sure of himself).

Mr. Babu: (Again stunned by Mr. B's comment) But, but, can you guarantee plating performance without knowing the pattern density?

B: …

Mr. Babu: Unbelievable. We have to provide comprehensive education on TEG…

It is not rare for tool manufacturers to encounter cases similar to the above. Knowledge of semiconductor devices and related technologies is essential.

Device TEG

Test Element Group (TEG): A TEG element for plating/polishing represents a certain region on a device. It is essential to determine specifications for plating and polishing performance.

See descriptions in the following chapters for details.

Necessary information for plating

Initial pattern profile and density are important.

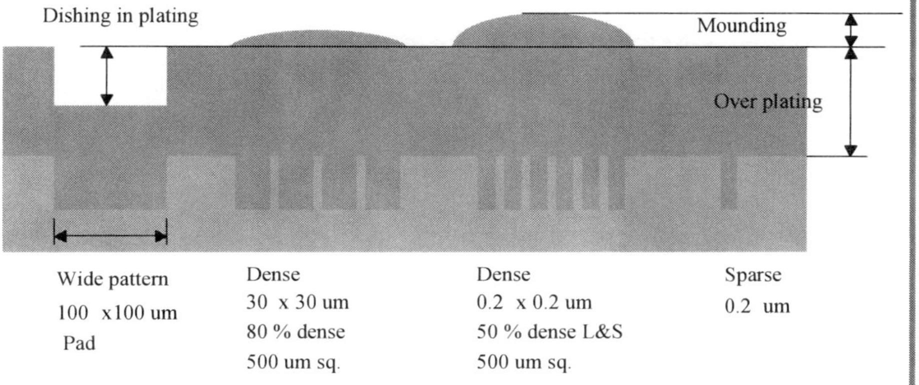

Necessary information for CMP

Initial pattern profile and density are important.

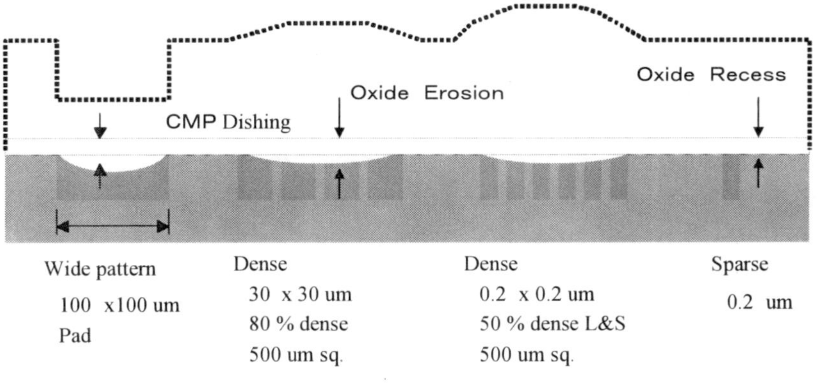

Let's study devices

Device face

Devices are mainly classified into memory devices and logic devices (see other technical books for details).

It may be seen that the examples of a memory device and a logic device totally differ from each other. The memory device has a fully packed memory area while the logic device has a variety of faces depending on the user's selection. Plating and polishing are conducted for each layer of wafers having a variety of faces.

Floor plan

Device manufacturing starts with an outline design called a floor plan.

- Making an overall plan (which module goes where).
- Making module plans (which block goes where).
- A cell is in blocks.

Concept of TEG

Let's study TEG. TEG has the following four steps.

Step 1: Electrical characteristics, such as the metal for BEOL development, open/short, and via-chain.

There are other optional TEGs, such as for CMP or for plating.

Step 2: FEOL/BEOL: For validation of logics.

Step 3: FEOL/BEOL: TEG wafer. The TEG area is set on the wafer.

Step 4: A commercially available wafer is detached from the TEG.

When you understand that the interconnect dimensions and design vary according to the user, device, and layer, and the presence of optional TEGs for the development of unit processes for plating or CMP, you will be ready for a demonstration with the data of plating or polishing.

You may also use a TEG wafer designed for a certain purpose made by a TEG specialized manufacturer.

You should recognize difference of faces of logic and memory.

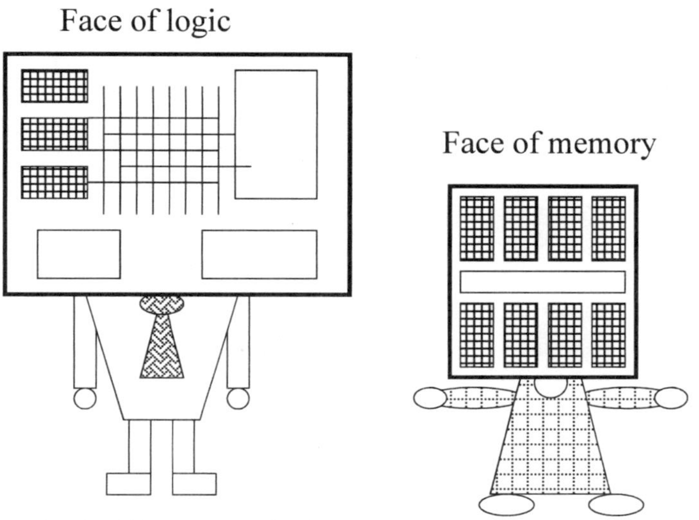

Floor plan (looks like a mansion)

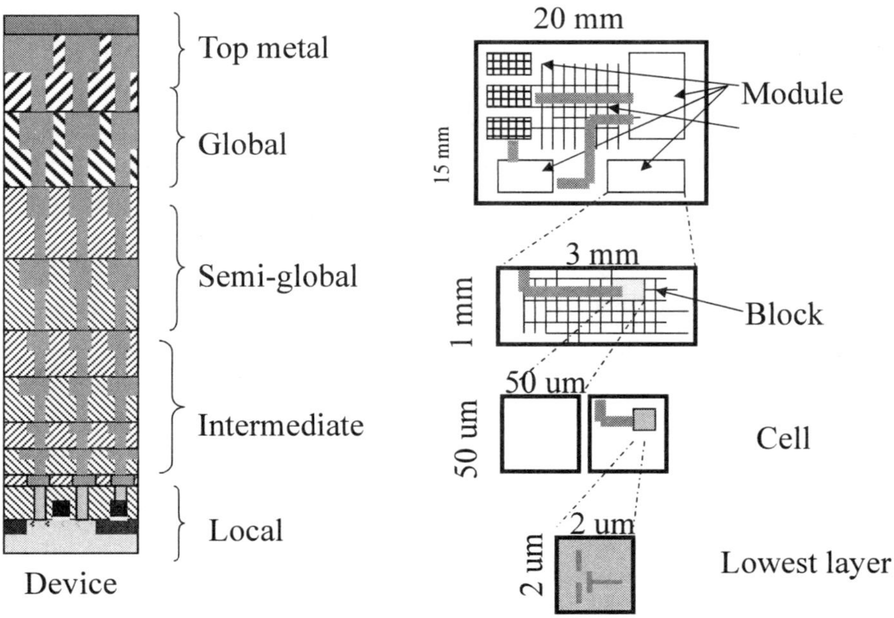

Bell Boy completed!

Scene 7: Order!

Finally, the order for the tools is placed with BER. BER struggles to develop the systems for many processes, including lithography, ion implantation, deposition, etching, CMP, plating, exhaust systems, and utilities, but the company manages to meet the delivery deadline.

Now, is the project successfully completed?
Far from it. SoC has hidden problems!
For details, see descriptions in the following chapters related to the multilayer interconnect problems. It has many problems anyway.

Scene 8: Bell Boy!

The figure on the next page illustrates the multilayer interconnect problems.
Although details will be given in the following chapters, there are many problems, such as dielectric film full of voids and Cu interconnects full of scratches and corrosion. As a result, no reliability is demonstrated. Further development is the key to solving such problems. It is advised that achievements presented by universities and conferences be followed up to boost the development of the next generation devices.
The project, which began with the reading by Dr. Manabu, has finally come to its completion and has overcome many challenges. Has it been heard that Bell Boy, the developed robot, is now working fine for a famous hotel?

This is the end of the fictional story. It contains a variety of important principles. I recommend that the readers read it again many times. The story has been used as a material for a role-play class in the universities and has been favorably received.

Now, let's discuss technical issues from the next section.

Interconnect problems

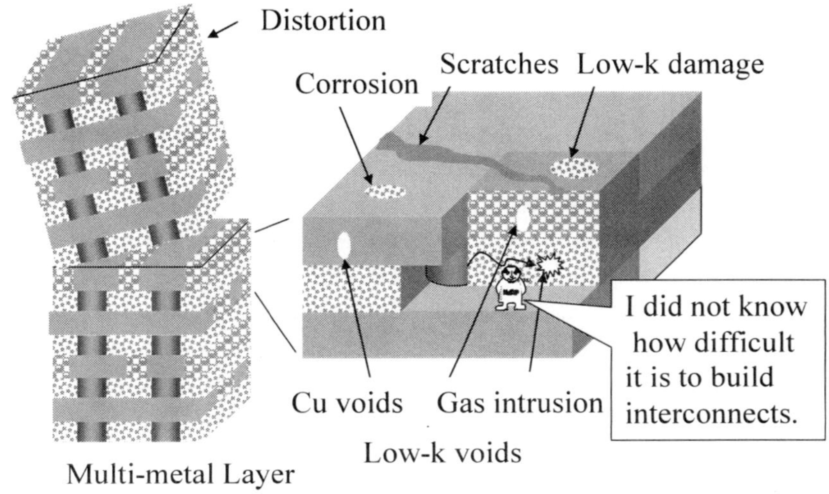

Completion of Robots

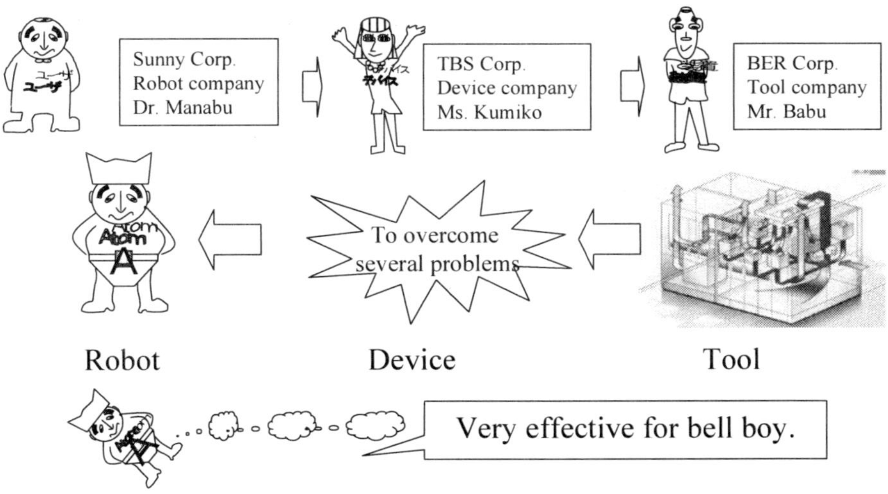

= The first dinner talk =

The first dinner talk was planned for the CAMP's 8th International Symposium on Chemical-Mechanical Planarization in 2003. About 100 engineers and their families attend the symposium every year. We thought that it would be nice to have an opportunity for the families to know what kind of work their parents or partners do and enjoy themselves at the same time. Prof. Babu appointed me to give the first dinner talk.

I talked about several subjects: that the semiconductor industry is like a *food chain* made up of chipset, device, tool (material) manufacturers; that the temper of Kumiko (my wife) goes up and down as the industry goes up and down; that multilayer interconnection is difficult like building a skyscraper and has many problems that must be addressed. I also introduced revolutionary (is it surely revolutionary?) CMP technologies, which would totally shock Dr. Preston, such as 0 psi pressure polishing and 10,000 rotation polishing.

When I finished the dinner talk dripping with sweat and stepped down from the podium, a little girl about 10 came and gave me words of appreciation, "Good job!" I felt relieved to know that she at least understood my English. Dinner talks are still given with the last one delivered at the 12th symposium in 2007. It seems that I am still the best speaker in terms of making the audience laugh.

Frequent Up and Down.
This is Bull-Whip Market!

Similarity between Interconnect and Skyscraper

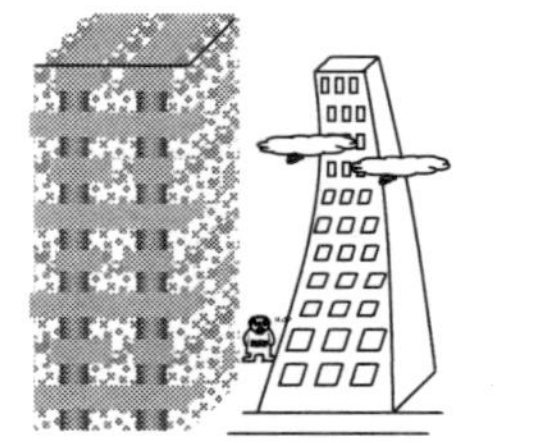

Magic hand process '0' psi CMP

Shinkansen Process '10,000 rpm CMP'

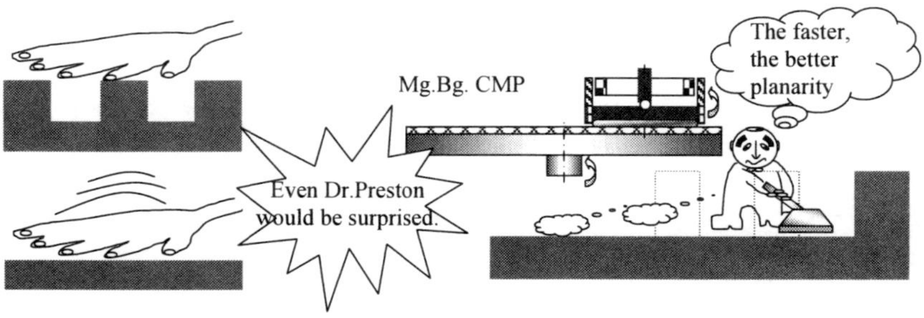

Chapter II

Wet Deposition Process : Plating

From plating in buckets to semiconductor plating

From the oldest plating to the latest plating

Plating, one of the oldest technologies in the world, was invented before Christ and came to Japan from China/Korea around the 4th century. I have continued to stress that the basics of plating development are found in buckets. Why? If you study the history and principle of plating, you will rush to a shop for buckets and stop making expensive equipment.

Look at the figure on the next page. As described later, Cu plating can be done only by applying a voltage of a few volts to a Cu plate and a target object in a large bath like a bucket. It is quite simple, isn't it? The basics of plating are simple, but its industrial application requires various additional components. For actual semiconductor plating systems, plating solution is circulated in a large bath with an anode (electrode) and a wafer (target object) placed in pair.

There are numerous variations of plating approaches. For example, stamp plating systems, a type of variation of *brush plating*, use less plating solution and require no sealing, which is the most troublesome for plating. They are based on the concept of forming film only at required locations. This concept has been further advanced to the development of *dummy plating systems* (described later).

Overview of the history of semiconductor plating technology

The history of semiconductor plating technology is outlined below. The history of plating technology for semiconductor manufacturing can be divided into two phases in terms of eras and capabilities. The first phase was the late 1980s during which bump plating technology was developed and adopted. This process fills pores of about 40 μm square and 40 μm height with Au to form Au bumps (electrode); its deposition thickness is on the order of several tens of micrometers. The second phase was the late 1990s during which interconnect plating technology was developed and adopted. This process fills submicron pores and grooves (equivalent to the design rule size) with Cu to form Cu interconnects. The figure shows similarity between polishing and plating systems in term of development and adoption timing. These systems were called 3K technologies; they were first used standalone, then were provided with built-in cleaning subsystems, and now are used further with monitor subsystems. Because plating was adopted for semiconductor manufacturing later than polishing, it was more easily accepted by the industry.

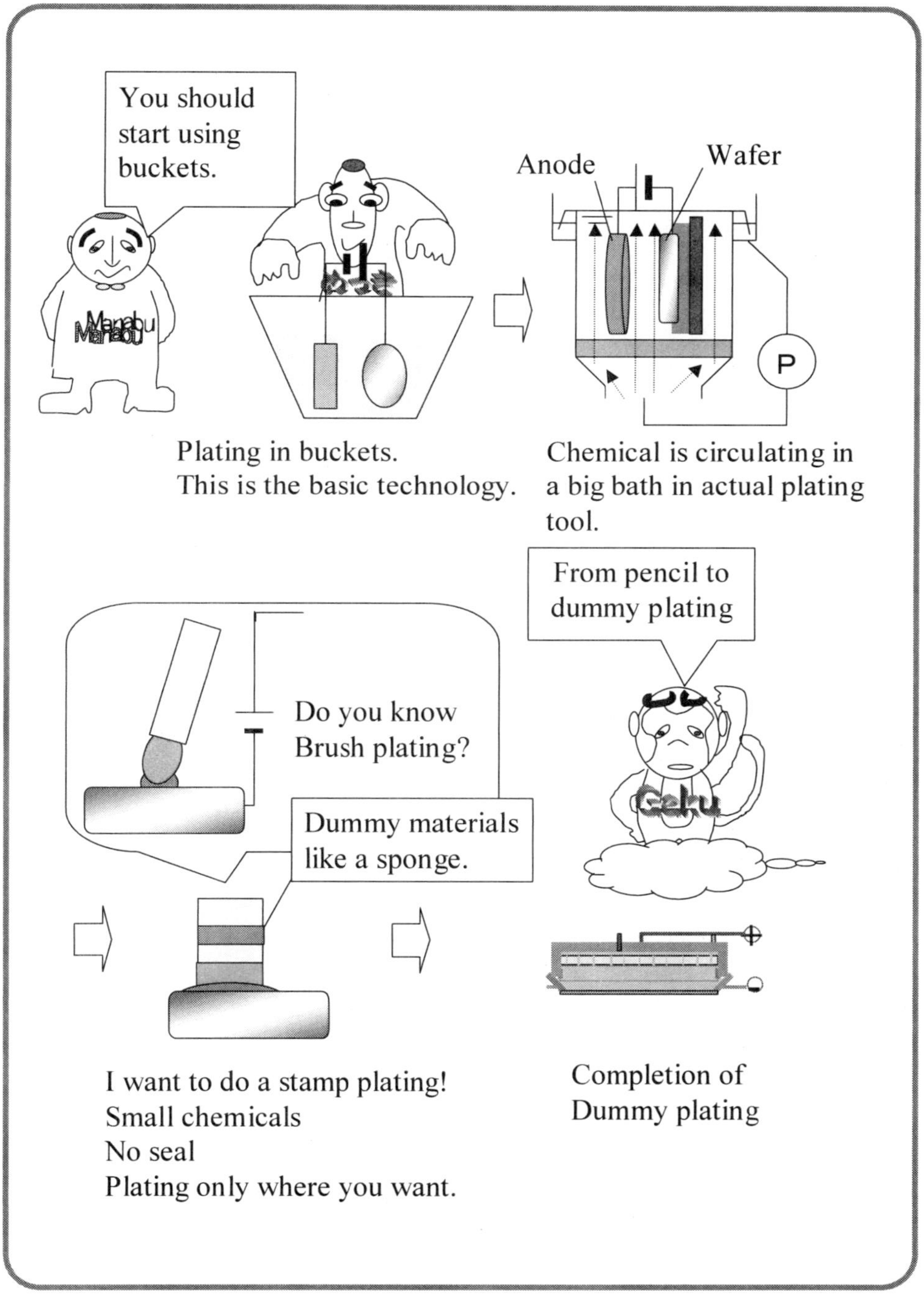
You should start using buckets.
Manabu
Anode
Wafer
P
Plating in buckets.
This is the basic technology.
Chemical is circulating in a big bath in actual plating tool.
From pencil to dummy plating
Do you know Brush plating?
Dummy materials like a sponge.
I want to do a stamp plating!
Small chemicals
No seal
Plating only where you want.
Completion of Dummy plating

Old *3K* to new *3K*

Bump plating for implementation

The feature of the Au bumping process is discussed below. Au bumping for semiconductor devices uses Au plating technology. Since plating was considered as a highly polluting process, Au bumping was conducted offline (outside clean rooms). It generated much dust around a reaction chamber and might contaminate wafers and the environment outside the system. Au bumping was a step near the end of the wafer manufacturing process, and wafer contamination was not a serious problem. However, contamination outside the system would be critical; it was impossible to install the Au bumping process in clean rooms used for front-end semiconductor device processes. At the same time, in-line processing (inside clean rooms) was much more desirable than offline processing to shorten the delivery time of semiconductor devices. An urgent task in the late 1980s was to transform traditional 3K plating technology to new 3K (kirei (clean), kantan (easy), and kagaku-teki (scientific)) technology. Thus, the development of plating process systems for semiconductor device manufacturing started. Its primary purpose was to develop clean systems that could be installed in clean rooms and to enhance the plating process so that bump pitch scaling required for the next generation could be achieved.

Interconnect plating

Interconnect plating is a main process in semiconductor device manufacturing. IBM announced at the end of 1997 that it would adopt Cu interconnects for logic devices and use plating technology for interconnect processing. This announcement stunned the industry again following the announcement of CMP. When I was first engaged in CMP development, it seemed to me that the adoption of polishing for semiconductor device manufacturing was impossible in terms of pursing ultrahigh vacuum *with no dust*. While wafer bump plating was already adopted in the late 1980s as described above, I did not think that plating could be employed for the interconnect process. I merely thought that the application of bumping to peripheral technologies should be established in preparation for the age of interconnect plating. For Cu interconnect process systems, the requirements of contamination control and high performance for scaling are more severe than those for bump process systems.

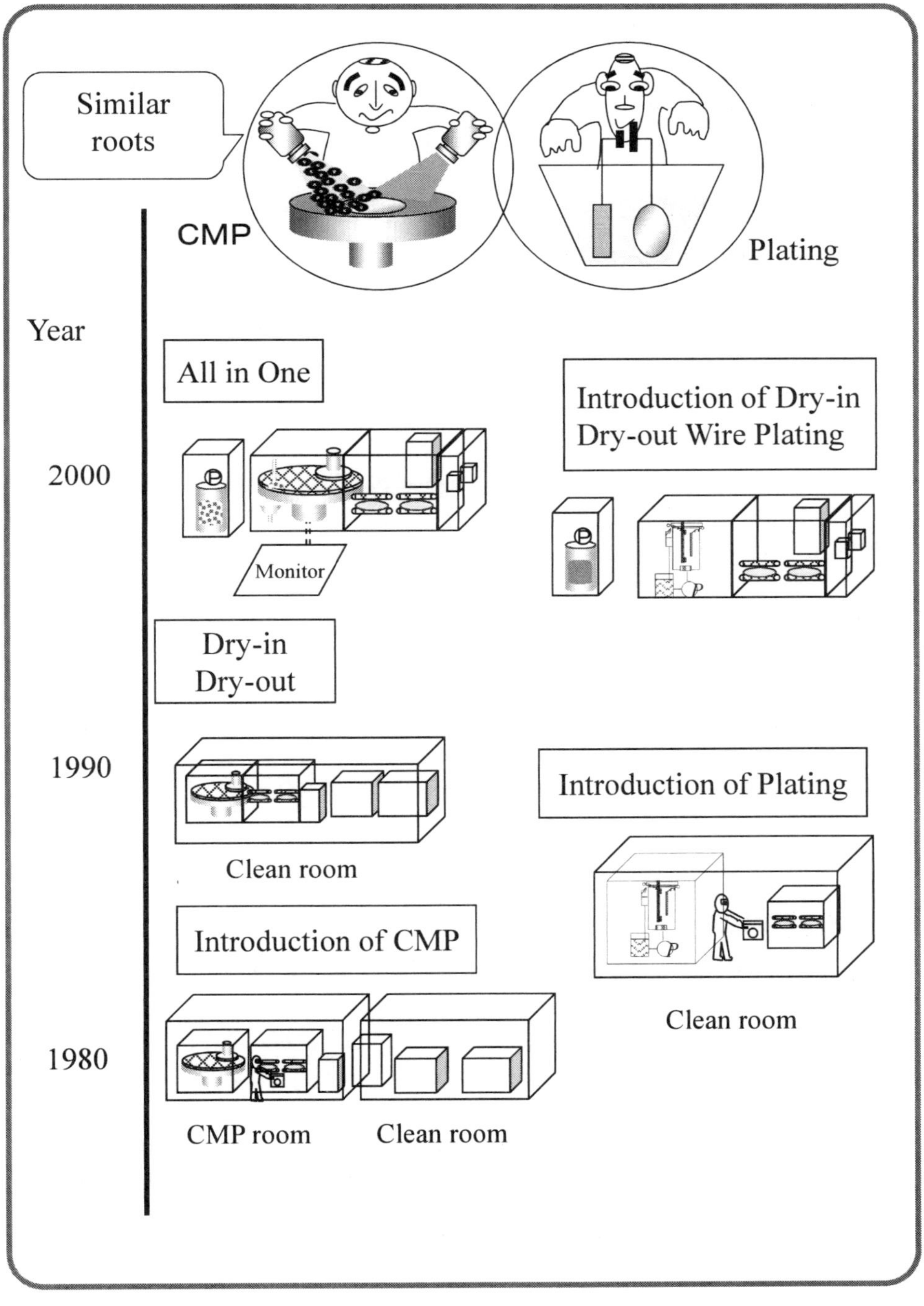

Similar roots
CMP
Plating
Year
All in One
Introduction of Dry-in Dry-out Wire Plating
2000
Monitor
Dry-in Dry-out
1990
Introduction of Plating
Clean room
Introduction of CMP
Clean room
1980
CMP room
Clean room

Various plating systems

Let's see the history of various plating systems. The figure on the next page shows the transition of plating system development.

Vertical dip method

Au bump plating for LCD driver IC was conducted using the so-called vertical dip method in the 1980s. This method covers the back surface of wafers with wax and dips them into a plating bath. Due to the difficulty of wax removal after plating, this method was replaced by the cup upstream method.

Cup upstream method

This method sets wafers with their top surface facing downward and blows plating solution from beneath the wafers. It can prevent deposition on the back surface of wafers since plating solution does not reach the back surface. This method, however, causes the plating of the wafer edge, and the plated Au is wasted in the subsequent chip dicing process. There is also a disadvantage that bubbles in plating solution remain on the wafer surface, causing plating defects.

New vertical dip method

As a solution to the disadvantages above, the new vertical dip method was developed. It covers the back and edge surfaces of wafers with a special jig and conducts the conventional vertical dip plating.

New cup upstream method

To avoid the problem for the conventional cup upstream method, the new cup upstream method performs the plating of wafers while rotating them for bubble removal. This method can reduce the adhesion of bubbles in plating solution to the wafer surface.

Thus, plating technology, one of the oldest technologies in the world, has been established as a cutting-edge semiconductor manufacturing process technology.

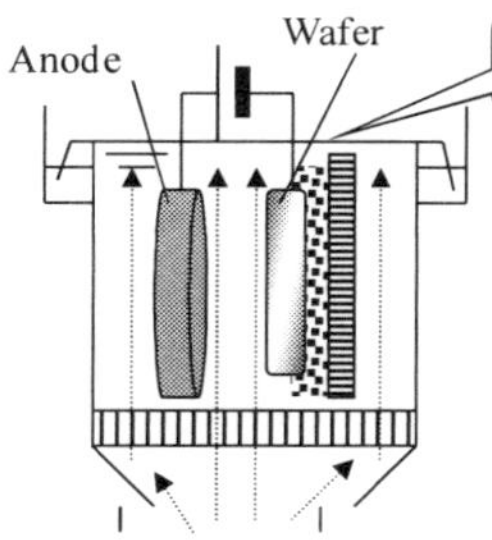

Vertical dip method

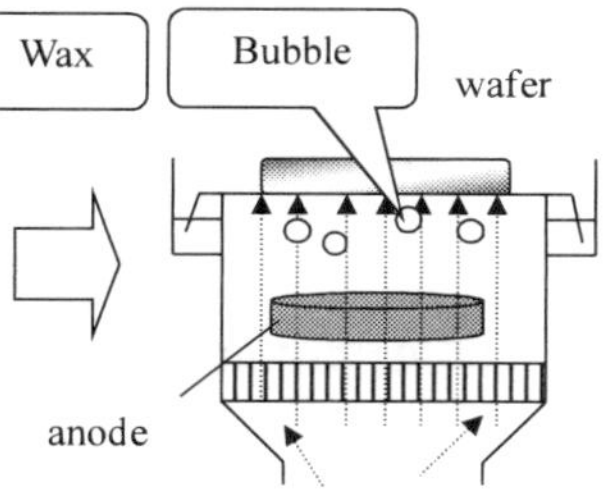

Cup upstream method

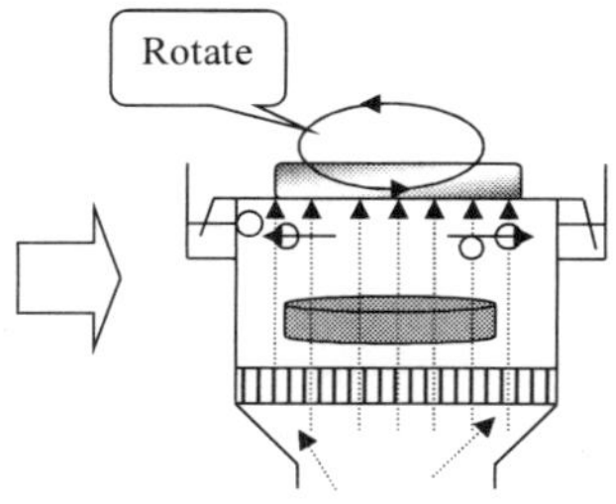

New Cup upstream method

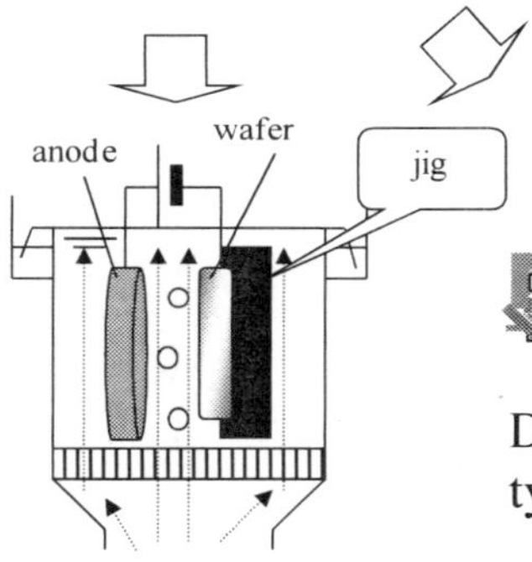

New vertical dip method

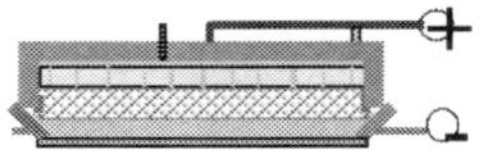

Dummy resistance type

TAB bump — Wire — CSP bump

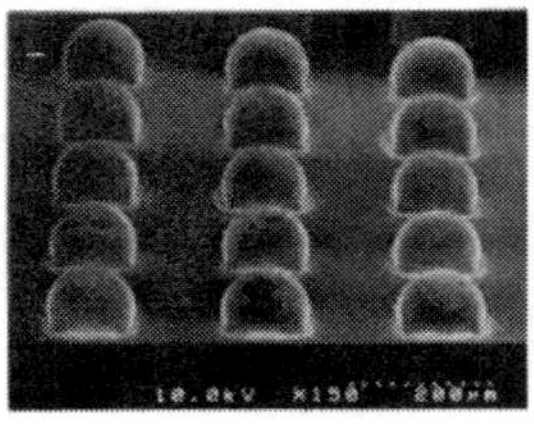

Bump

Wire gap filling
0.2μm dia. AR=4.5

Four applications

There are four major applications of semiconductor device plating. The details of each application will be described later.

Bump plating (Au, etc.)

In the final step of the multilayer interconnection process for semiconductor devices, an electrode (bump) for external connection (e.g. interposer) is made by Au plating. The size and number of bumps depend on the device type. In the 1980s, the bump size was expected to scale to the order of dozen micrometers, and the number of bumps was expected to increase to several thousands. The optimal plating technology for scaling was immediately developed and adopted.

Interconnect plating (Cu)

In the 1980s, a process of forming Cu interconnects through plating and polishing, the so-called damascene process, was invented. This process etches dielectric film to make interconnect grooves. In the grooves, it forms barrier and seed layers by sputtering and then Cu interconnects by plating. Finally, it performs CMP to finish the creation of interconnects.

Cap metal plating on interconnects (Co-W, etc.)

Since Cu multilayer interconnects was adopted, plating technology has been applied in various stages of semiconductor manufacturing. An example of such cases is the appearance of deposition technology that forms film after Cu interconnect plating like putting a cap. This technology is called cap metal plating on interconnects. It aims to improve the reliability of interconnection, as detailed later.

Re-distribution interconnect plating (Cu, Ni, etc.)

Re-distribution interconnect plating, which changes the interconnect position for external device connection, has also employed the plating of Cu or Ni.

In addition to the four applications above, the range of the application of plating technology, or wet deposition process, will definitely be further enriched in the future.

Four(4) important application

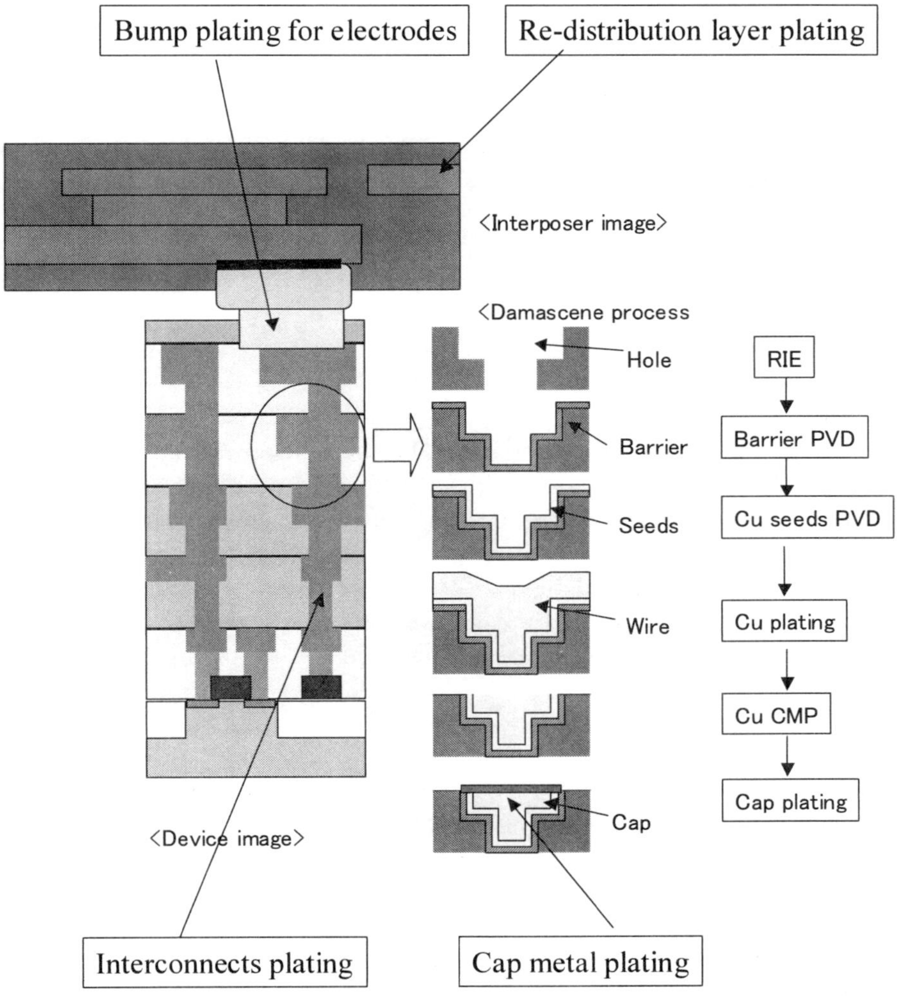

Six subsystems

Subsystems for the plating process include (1) chemical supply and control, (2) plating, (3) cleaning, (4) transfer, (5) load/unload, and (6) waste treatment. This wet process configuration is similar to that of CMP systems.

Chemical supply and control

The type of plating solution depends on the user's decision. Various parameters, such as plating solution concentration, chemical agent, temperature, pressure, and filtration, must be controlled. A plating solution supply system provides the required control, as well as supplying plating solution.

Plating

Various methods of plating are available. Plating performance is determined by the know-how of each manufacturer. See the appropriate description for details.

Cleaning

Cleaning and drying are essential for the introduction of dry-in/dry-out wet processes although their specifications vary according to the plating process. Cleaning is also intended to perform surface modification after plating.

Transfer and load/unload

For transfer, a common platform for wet processes can be used. In fact, the plating system developed by Ebara Corporation has adopted the same transfer and load/unload platforms as those of CMP systems.

Waste treatment

Waste treatment is required for plating. Especially, environmentally friendly design for plating solution recycling/reuse will be increasingly demanded.

Dry-in/dry-out

Dry-in/dry-out technology has been essential for the wide spreading of plating systems. Especially for 300 mm wafer systems, it has been developed as mini-environment technology.

Total coordination

The building of comprehensive systems, each of which covers plating solution supply, deposition, cleaning, and waste treatment, is demanded.

The figure on the next page shows an overview and appearance of a plating system. The appearance of the system is the same as that of other dry systems.

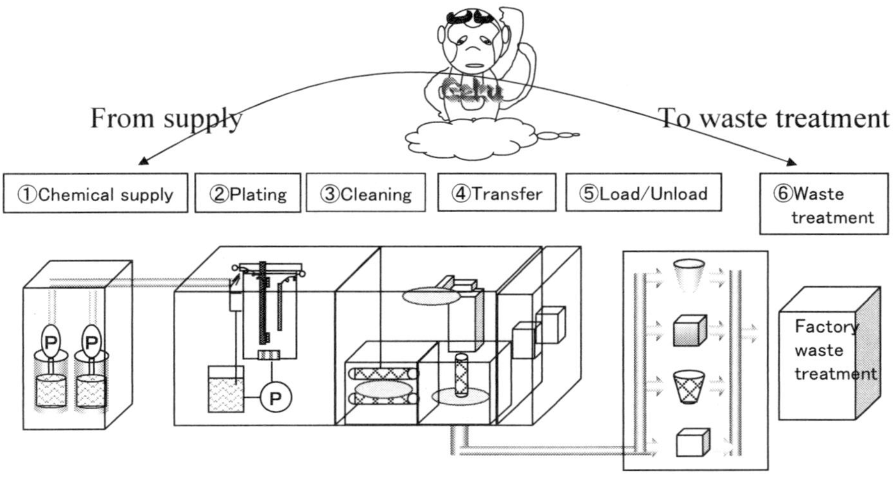

Sub-systems for plating

Photo of Plating Tool

Let's start with Faraday's Law

Faraday's Law is always mentioned when plating is discussed. Its basic knowledge is explained below. The understanding of Faraday's Law is indispensable to plating, as well as that of Preston's Law for polishing.

Soluble/insoluble anode and reaction equation

There are two types of anodes: soluble anode and insoluble anode. For distinguishing the detail between them, see other technical books. In this section, the minimum essential information required for semiconductor plating technology is given with $CuSO_4$ plating taken as an example.

For a soluble anode, the reaction follows the equation below. The voltage to be applied is lower than the electrolytic reaction voltage of water (1.25 V or less).

Ionic dissociation in aqueous solution: $CuSO_4$ -> Cu^{2+} + SO_4^{2-}

Anode (+) reaction: Cu -> Cu^{2+} + 2 e^-

Cathode (-) reaction: Cu^{2+} + 2 e^- -> Cu

For an insoluble anode, the reaction follows the equation below.

Ionic dissociation in aqueous solution: $CuSO_4$ -> Cu^{2+} + SO_4^{2-}

Anode (+) reaction: 2 OH^- -> O_2 + H_2 + 2 e^-

Cathode (-) reaction: Cu^{2+} + 2 e^- -> Cu

Application of Faraday's Law

Let's calculate the thickness of Cu deposition on semiconductor devices. Based on Faraday's Law, the volume of Cu deposition is proportional to the coulomb force. The equation below can be used to easily calculate the rate and volume of Cu deposition on wafers. It is recommended to remember this useful equation. The calculation is based on a current density generally used for plating (20 mA/cm^2).

Volume of plating deposition based on Faraday's Law = (Current value) x (Plating time) x (1 gram equivalent of Cu)

Thus, the deposition rate of plating film is calculated to be 0.44 μm/min (with a current density of 20 mA/cm^2).

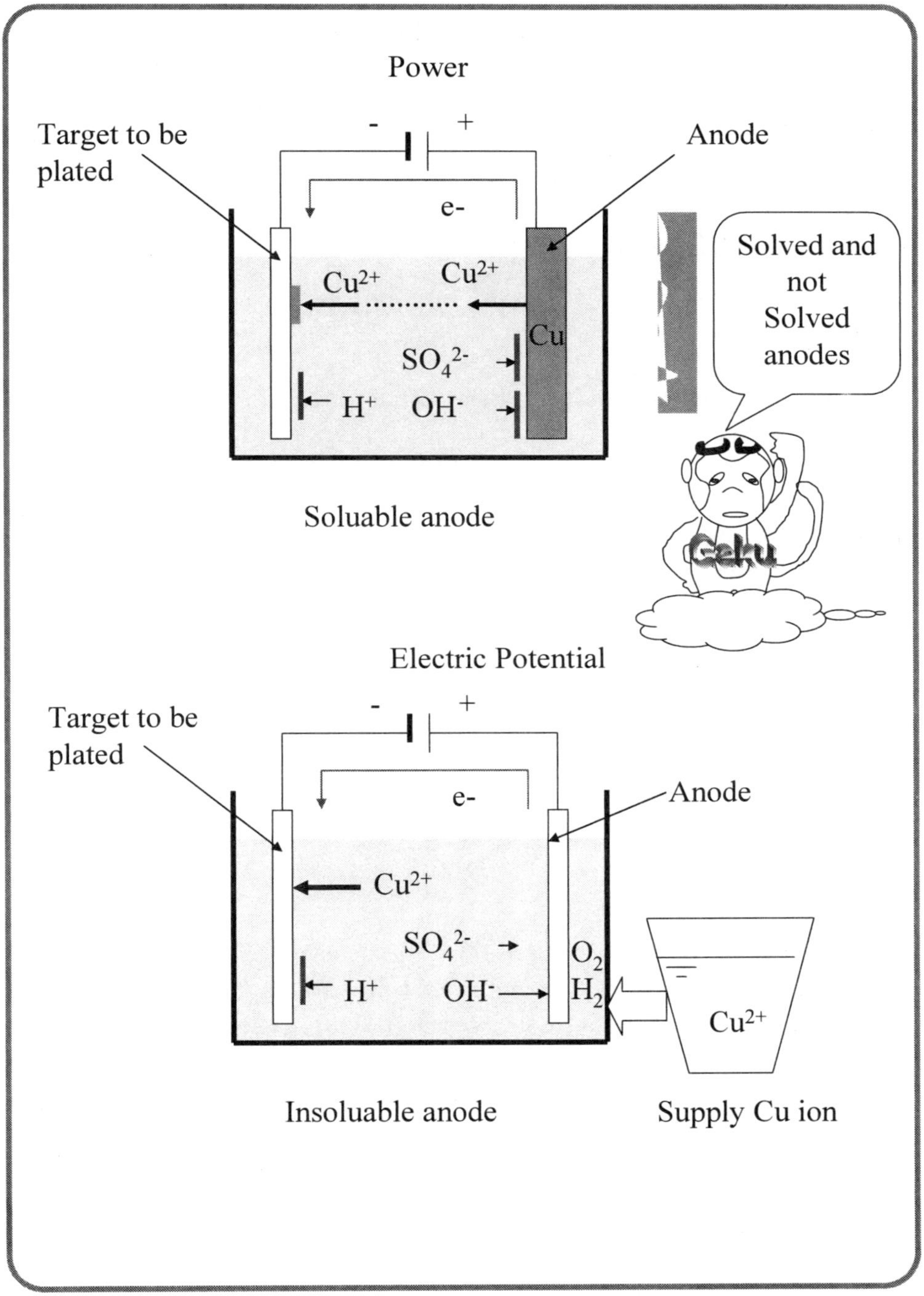
Power
- +
Target to be plated
Anode
e-
Cu^{2+}
Cu^{2+}
Cu
SO_4^{2-}
H^+
OH^-
Solved and not Solved anodes
Soluable anode
Electric Potential
- +
Target to be plated
Anode
e-
Cu^{2+}
SO_4^{2-}
H^+
OH^-
O_2
H_2
Cu^{2+}
Insoluable anode
Supply Cu ion

Three secret drugs

The simple description of the roles of chemical agents is given below. The plating process depends on know-how related to chemical agents. It may not be an exaggeration to say that chemical agents determine the entire performance of plating. Chemical agents are basically categorized into (1) suppressors, (2) accelerators, and (3) levelers. The role of each agent is explained in this section.

Suppressor

A suppressor adheres to the Cu surface with the mediation of chlorine. Then, plating deposition on the Cu surface is suppressed, locally reducing the deposition rate and improving the nucleation density. The difference in the volume of initial adhesion between the field and the trench bottom (adhesion on the field is less than that on the trench bottom) may affect bottom-up filling. A surfactant is often used as a suppressor.

Accelerator

An accelerator, competitive to a suppressor in terms of adhesion, decreases the plating deposition potential. Staying on the plating surface, it has a locally high concentration due to the unevenness of the plating surface, further accelerating the plating process. In other words, it accelerates plating in concaves. An organic sulfur compound is a typical example of an accelerator.

Leveler

A leveler suppresses plating deposition. Since adhesion depends on solution stirring, the adhesion is difficult in areas where solution flow is slow, and the diffusion layer is thick, such as the trench bottom. A leveler suppresses the deposition acceleration effect of an accelerator and improves the evenness of film thickness. A quaternized amine compound is a typical example of a leveler.

The categorization of chemical agents is roughly summarized above. See other technical books for details on plating. These three types of chemical agents may be categorized into two types in some cases.

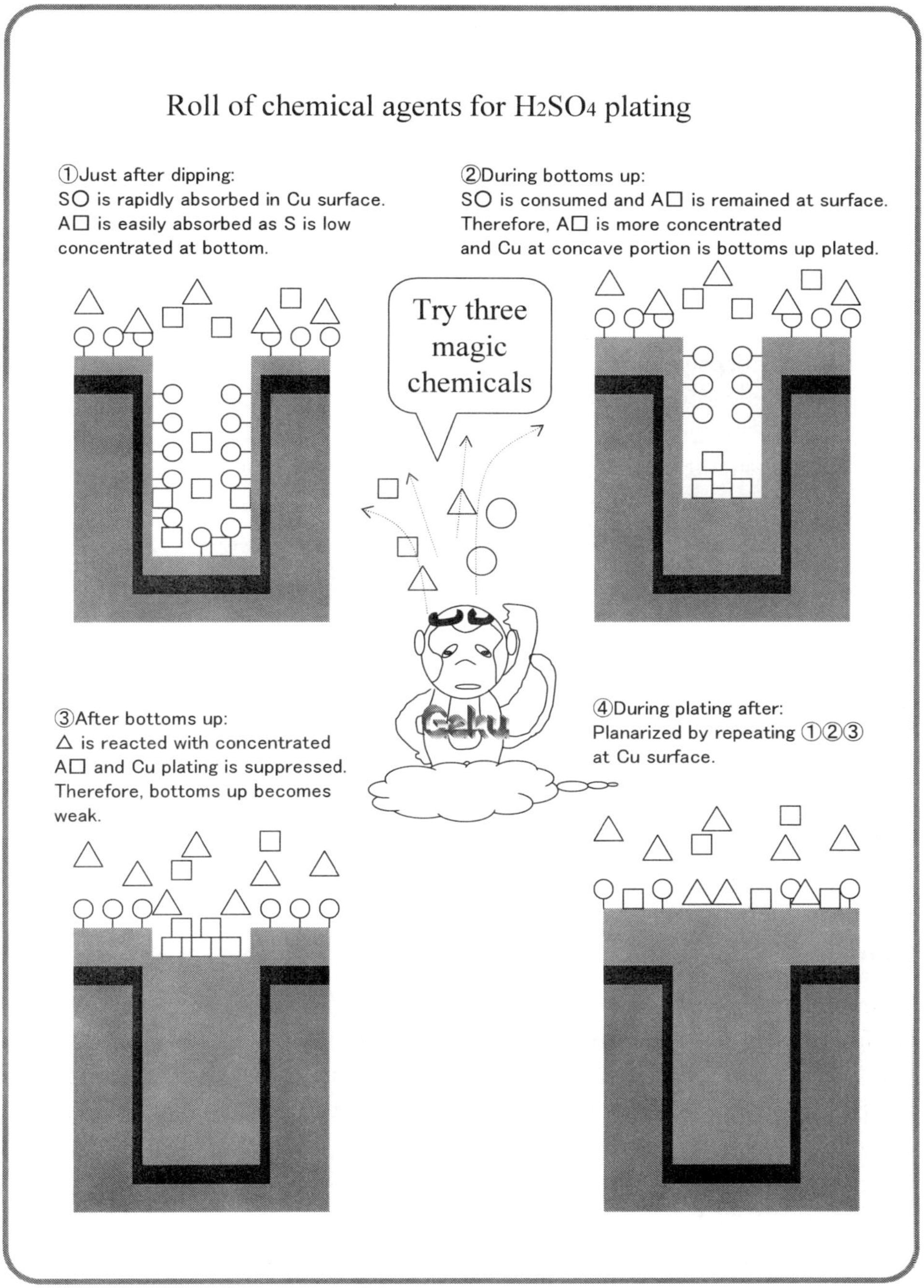
Roll of chemical agents for H_2SO_4 plating
①Just after dipping:
SO is rapidly absorbed in Cu surface.
A□ is easily absorbed as S is low
concentrated at bottom.
②During bottoms up:
SO is consumed and A□ is remained at surface.
Therefore, A□ is more concentrated
and Cu at concave portion is bottoms up plated.
Try three magic chemicals
Geku
③After bottoms up:
△ is reacted with concentrated
A□ and Cu plating is suppressed.
Therefore, bottoms up becomes
weak.
④During plating after:
Planarized by repeating ①②③
at Cu surface.

Plating performance requirements

Plating performance

Plating performance is evaluated by (1) deposition rate, (2) wafer level non-uniformity, (3) defects, (4) gap filling capability, (5) adhesion, (6) resistivity, and (7) contamination.

(1) The deposition rate follows Faraday's Law as the function of current density and metal ion concentration in plating solution.
(2) The wafer level non-uniformity is achieved by providing a uniform current density across the wafer surface. To do so, it is required to ensure a uniform flow of plating solution on the wafer surface and to provide electric-field control.
(3) Defects are determined by the sum of numbers of voids, foreign materials, corrosions, and erosions.
(4) The gap filling capability is adjusted by chemical agents. Its improvement is required because gap filling will be further difficult as scaling proceeds.
(5) Adhesion with barrier film is an important parameter that affects electromigration (EM) performance. Of course, delamination is a hopeless case.
(6) Low resistivity is one of purposes of Cu interconnection. A resistivity of 2.2 μΩcm is desirable for interconnects compared to 1.6 μΩcm for the bulk.
(7) For cleanness, impurities should thoroughly be removed. However, controlled impurities are required to enhance EM resistance.

Example

An example of gap filling with typical $CuSO_4$ plating is given with current density and Cu ion concentration taken as parameters. Generally, an increase in the deposition rate with an increased current density or Cu ion concentration degrades the wafer level uniformity. Plating conditions must be decided considering the uniformity.

Visualization of flow

The uniformity of solution flow provides a basis for the evaluation above. To achieve a uniform flow, a hydraulic passage design has been developed based on computer analysis and flow visualization.

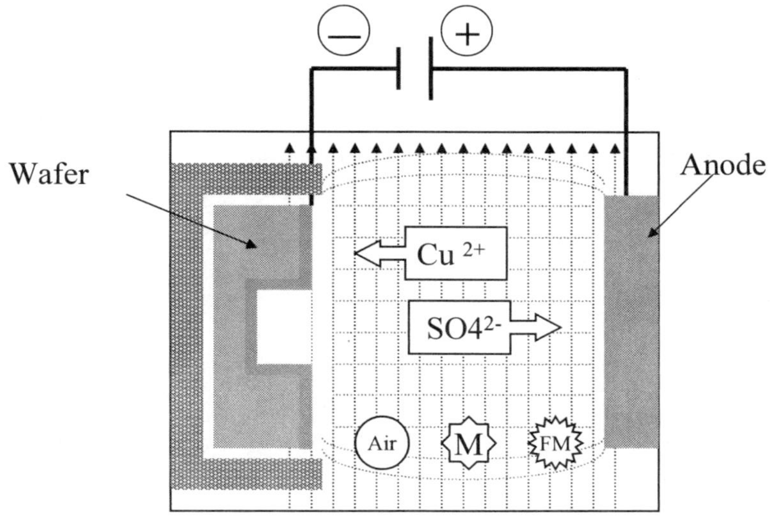

Principle of plating

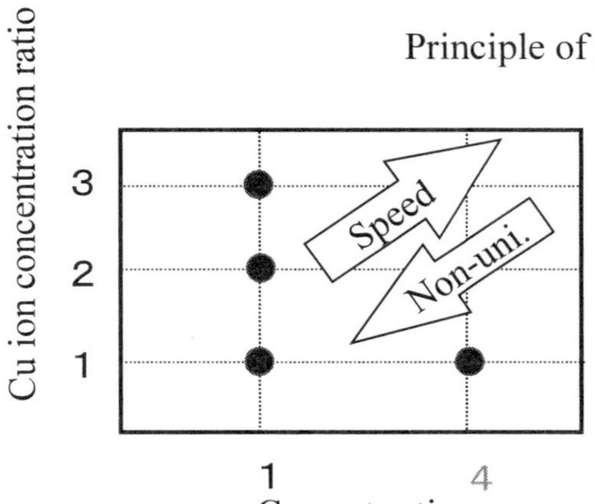

Non-uniformity and deposition speed

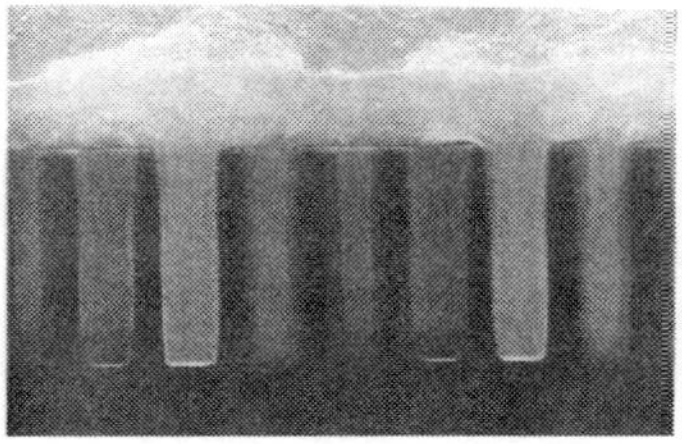

Gap filling

Cleaning and contamination control

Cleaning

The cleaning surface is made of various materials (Si, SiO_2, Si-N, Cu, Ta, Ta-N, etc.) used for semiconductor processes, and removal targets are plating solution and foreign materials (FM). It is required to sufficiently understand device processes when considering corrosion countermeasures for the base material with the mixed presence of these materials.

The features of various cleaning modules are explained below.

(1) Roll type: A standard system that rotates roll members with several hundred rpm to clean the front and back surfaces of wafers at the same time.

(2) Pen type: A standard system, similar to (1), that rotates pen members with several hundred rpm to clean the front surface of wafers.

(3) Jet type: In addition to the standard jet type, jet cleaning with ultrasonic waves and cavitation-based special jet cleaning are available. The cavitation jet provides a wider range and higher efficiency of cleaning than those of the conventional type. The differences in the cleaning range and efficiency between the conventional type and the cavitation jet type are doubled for wafers with uneven surfaces, instead of blanket wafers.

(4) Spin drying: This process is installed in the pen type module (2) or the jet type module (3) for final drying.

Contamination control

Cu was not welcomed for semiconductor devices, and introducing Cu plating or CMP into clean rooms was controversial. Cu-related systems largely affect the Cu contamination control design of clean rooms. After trials and errors, it was found that safety was secured by installing all Cu-related systems into a special room, but its cost was high; installing such systems into a general clean room might be cost-effective, but it imposed a higher risk of contamination. Compromise approaches were also proposed. The selection of approaches depends on the engineering of device manufacturers. Anyway, systems should be provided with proper airflow design to avoid external contamination. The airflow design of CMP systems has been verified.

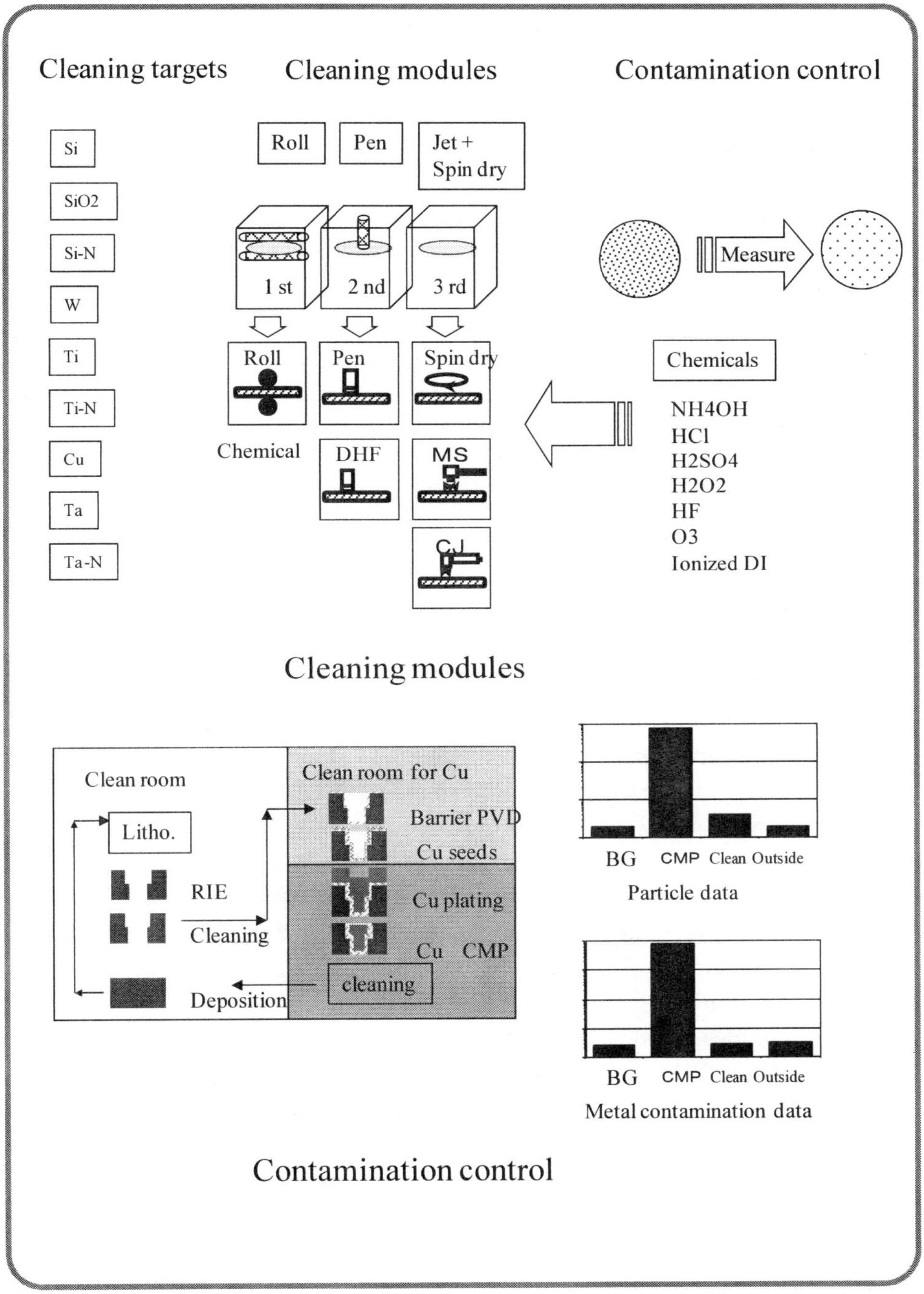
Cleaning targets
Cleaning modules
Contamination control
Si
SiO2
Si-N
W
Ti
Ti-N
Cu
Ta
Ta-N
Roll
Pen
Jet +
Spin dry
1 st
2 nd
3 rd
Roll
Pen
Spin dry
Chemical
DHF
MS
CJ
Measure
Chemicals
NH4OH
HCl
H2SO4
H2O2
HF
O3
Ionized DI
Cleaning modules
Clean room
Litho.
RIE
Cleaning
Deposition
Clean room for Cu
Barrier PVD
Cu seeds
Cu plating
Cu CMP
cleaning
BG
CMP
Clean
Outside
Particle data
BG
CMP
Clean
Outside
Metal contamination data
Contamination control

Fill in small hole with pinch-off?

Demands for interconnect plating

To discuss possible interconnect requirements in the future, let's see the requirements for 2010 predicted in the International Technology Roadmap for Semiconductors (ITRS) 2005. The filling aspect ratio for the interconnect/via gap is 1.8/1.6. The actual aspect ratio for the interconnect/via gap will be 2.25/2.26 when considering a barrier thickness of 3.3 nm and a seed layer thickness of 3.3 nm. The rough surface on an oxide film etching sidewall before barrier formation and the shape of sputtering will provide a shape that is difficult to perform gap filling. Also, the resistivity of the seed layer of 3.3 nm may be extremely high, increasing radial non-uniformity for 300 mm wafers. If the seed layer is removed, the barrier serves as an electrode, resulting in further higher resistivity and increased radial non-uniformity for 300 mm wafers. In the trend of scaling, the 45 nm generation in 2010 may face the issues of a high resistivity of seed layer (including barrier seed layer) and a shape difficult for gap filling.

Plating with dummy material resistance

An innovative technology that solves the problems above is introduced below. If a seed layer gets thinner with the conventional plating method, in-wafer radial uniformity decreases. With a dummy material having the dummy resistance function, in-wafer radial uniformity does not decrease even if a conducting seed layer gets thinner, offering the world's highest level of plating performance at present (in-wafer non-uniformity of film thickness at $3\sigma = 10$ % and aspect ratio for contact hole gap filling with an interconnect width of 0.1 μm = 4.5).

Note) Conventional circulation bath plating (not using a dummy material having the dummy resistance function) can achieve the performance above (in-wafer non-uniformity of film thickness at $3\sigma = 10$ % and aspect ratio for contact hole gap filling with an interconnect width of 0.2 μm = 4.5) for a seed layer of 60 nm thickness, but not for that of 20 nm thickness.

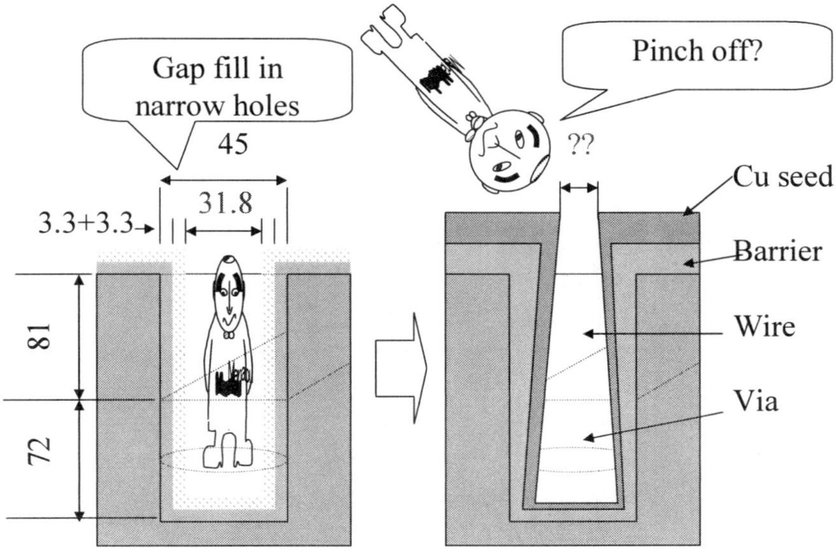

Wire and via dimensions for 45 nm device in 2010

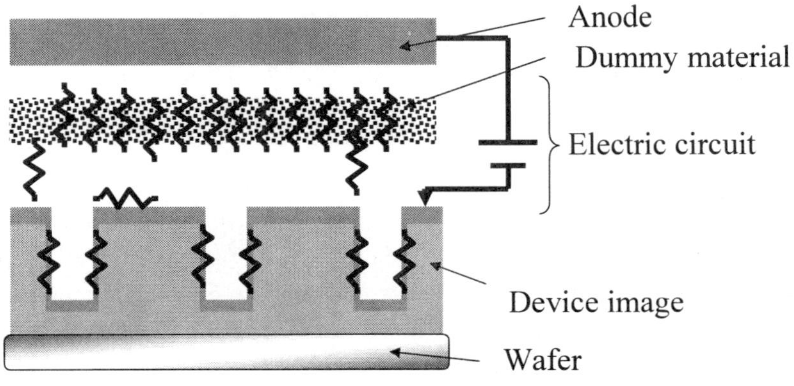

Dummy plating tool image

Only 60 cc bath?

System overview

The anode head structure and module structure for plating systems are described below.

(1) Anode head structure

A dry contact type of anode head structure is adopted, which has an electrode contact in the edge seal and sets wafers with their top surface facing upward. Plating solution is placed at a narrow pitch of 1 mm or less on each wafer. A porous ceramic is set between the anode and a wafer to provide dummy resistance. Plating solution can be supplied through the inlet to the upper and lower chambers.

(2) System module structure

The system module consists mainly of the anode head, stage, and anode standby section. An example of the operation sequence is shown below.

- A wafer is loaded at the lower position.
- The stage moves to the upper position for pretreatment and deposition.
- The stage moves to the middle position for cleaning and drying.
- The stage moves to the lower position for wafer unload.

The stage moves up/down for each step: wafer load/unload at the lower position, pretreatment and cleaning/drying at the middle position, and plating at the upper position. A tray filled with plating solution is set at the anode standby section, and the anode at the standby position remains dipped in plating solution. The anode head consists of high phosphorous Cu and dummy material (porous ceramic, etc.). The dummy material protects black film on the anode surface, prevents plating solution from spilling during the anode travel, and controls the electric field. $CuSO_4$ is used as plating solution, and its volume of use is small (60 cc) because of a narrow pitch. The plating solution is fed between a wafer and the anode head immediately before plating; it is recovered through nozzles after plating.

Key for dummy material: porosity

The dummy material used to provide dummy resistance is a porous ceramic like pumice. It limits the passage rate of plating solution to increase apparent resistivity.

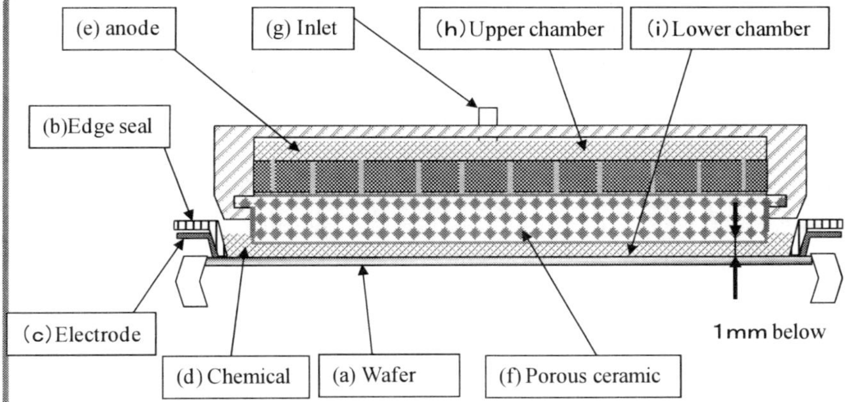

Dummy plating head

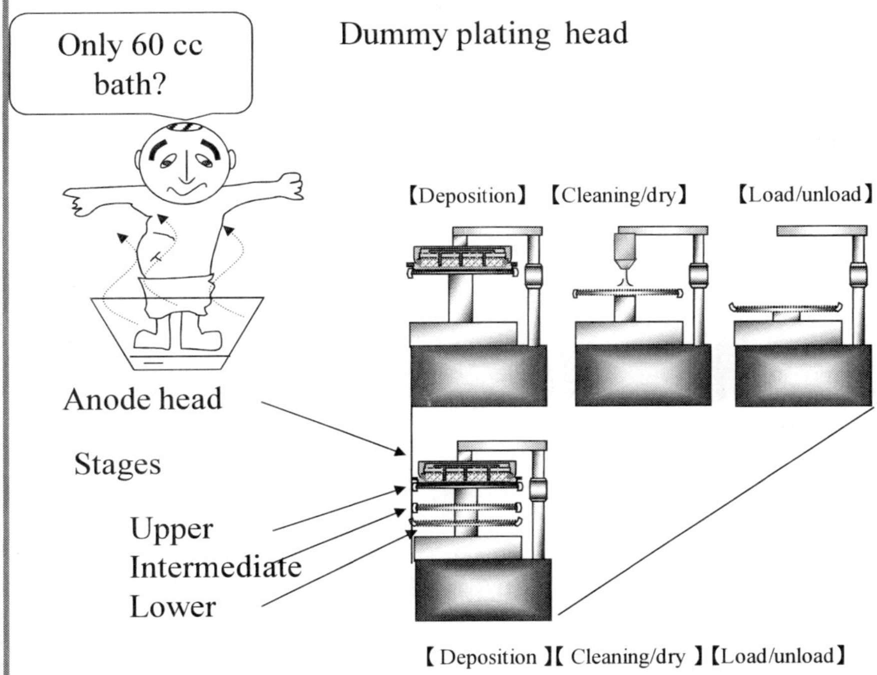

Function of each module

Dummy plating tool description

FEM (finite element method) in any case

Analysis of plating solution flow and electric field distribution

The plating deposition rate is mainly determined by the current density and metal ion concentration of plating solution. Deposition non-uniformity can be controlled according to plating solution flow and electric field distribution. The difference in these two factors between narrow-pitch static bath plating, such as dummy plating, and conventional circulation bath plating is explained below.

	Conventional circulation bath plating	Narrow-pitch static bath plating
Flow	Controlled to make uniform flow	Being static, controlled to make completely uniform flow
Electric Field	Controlled according to the thickness	Controlled according to the thickness

Conventional circulation bath plating requires controlling the flow of plating solution circulated between a wafer and the anode, as well as the electric field. On the other hand, in narrow-pitch static bath plating, the flow field is completely uniform on the deposition surface of a wafer. To control deposition distribution for narrow-pitch static bath plating, only the control of the electric field is required.

Analysis of electric field control for narrow-pitch static electro-chemical deposition (ECD)

The figure on the next page shows the result of analyzing the electric field of narrow-pitch static ECD with FEM.

Analysis conditions: Pre/post-processor: FEMAP

Solver: ASTEAN

Two-dimensional axially symmetric model

The figure indicates that the potential near the wafer is uniform.

The dummy material is like pumice as mentioned on the previous page, and its resistance increases as its porosity decreases. The analysis result shows that the terminal effect is reduced by decreasing the porosity. However, lower porosity also disrupts the flow of plating solution. The optimization of this trade-off relationship is critical, and FEM is greatly useful for this purpose.

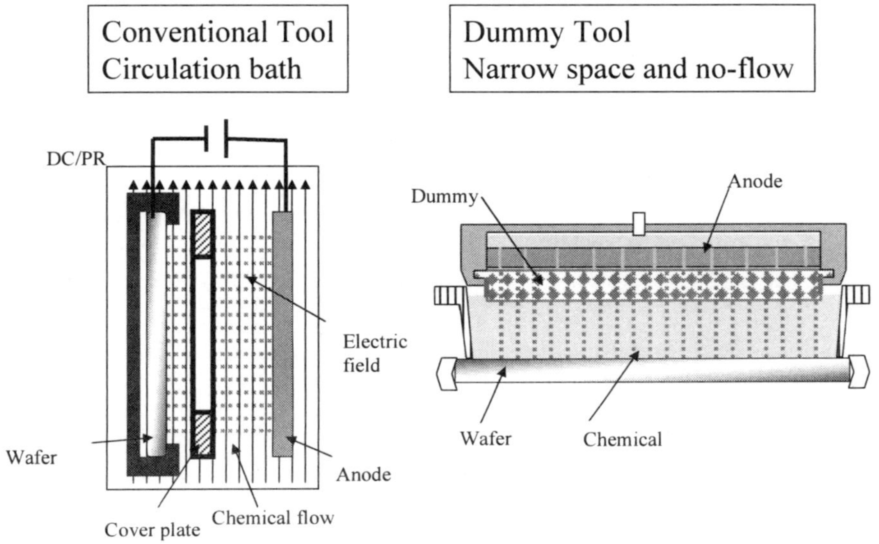

Difference between conventional and Dummy plating tools

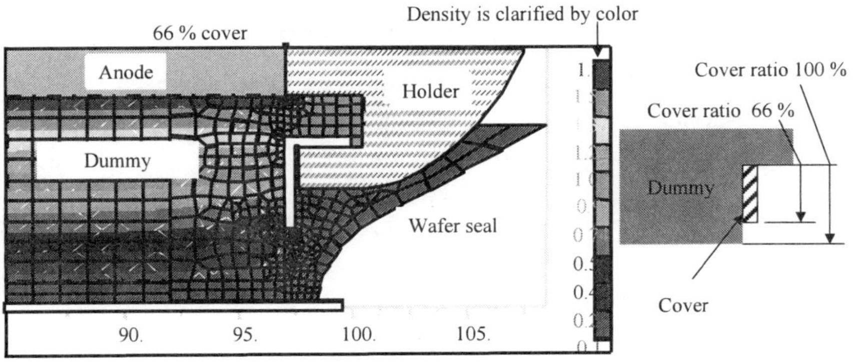

Electric field simulation by FEM

Dummy material for terminal effect minimization

Effect of the conducting seed layer thickness (terminal effect)

A thinner barrier layer is accompanied by a thinner seed layer. The electric resistivity of a seed layer depends on the radial distance on a wafer. When the seed layer thickness decreases, a change in this radial distance affects the deposition uniformity. This is called *terminal effect*, which is a major concern in fine pitch interconnect plating. For a thinner seed layer formed with the conventional plating method, the plating thickness becomes thinner toward the center. On the other hand, dummy plating with dummy resistance can provide uniform deposition for a thinner seed layer, just as in the case of thick film.

Film thickness control

Film thickness can be controlled by changing the porosity of dummy resistance material or by masking the material edge. As shown in the figure on the next page, a surface profile suitable for the subsequent CMP process can be obtained by increasing or decreasing the thickness of the edge of dummy resistance material.

Effects of dummy plating with dummy resistance

The effects of dummy plating can be summarized as follows.

(1) Conventional circulation bath plating (using no dummy material with the dummy resistance function) gives lower uniformity when the seed layer thickness is reduced to about 20 nm or less.

(2) A dummy material with the dummy resistance function allows forming a thinner conducting seed layer.

(3) A dummy material also allows film thickness control.

(4) The in-wafer non-uniformity of film thickness is $3\sigma 3$ % or less with 3 mm edge exclusion (EE).

A plating system based on the concept of narrow-pitch static bath plating has been developed by adopting a dummy material with the dummy resistance function. The system has achieved the world's highest class plating performance and demonstrated that this plating method is applicable to thinner conducting seed layers (a thinner barrier layer is accompanied by a thinner seed layer). By this method, the possibility of applying plating technology to Cu interconnection on seed layers for the 32 nm generation or beyond has been suggested.

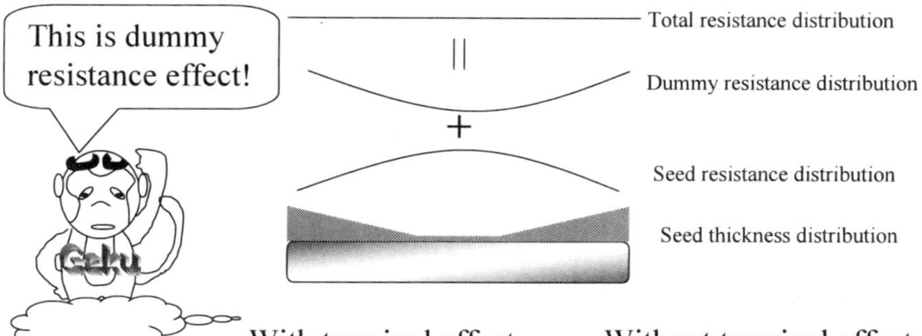

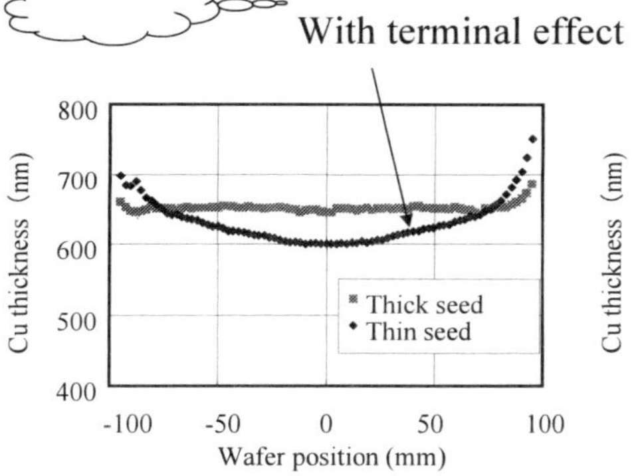

Conventional plating

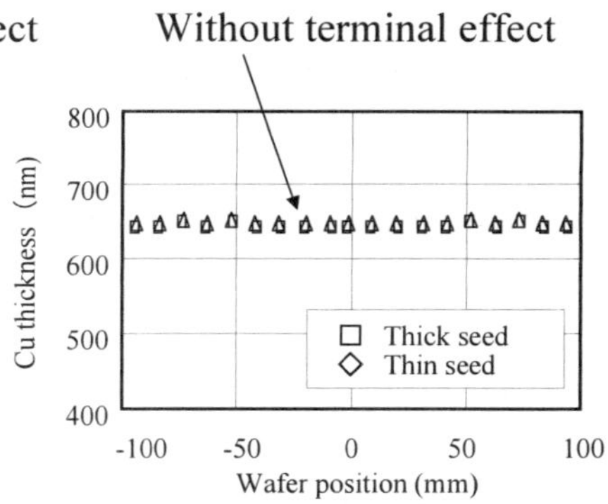

Dummy plating

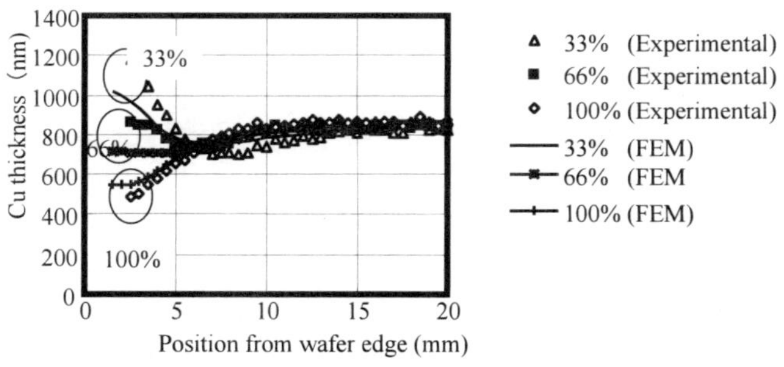

Thickness control at wafer edge

Porcupine process

Planar deposition

The history of deposition has coincided with the history of planarization. Demands for gap filling conflict with demands for the formation of planar film. Conformal deposition always makes a non-uniform surface. Since plating allows bottom-up filling as described earlier, both gap filling and planar film can be achieved at the same time for microholes. However, it is hardly possible if a substrate also has a wide pattern.

The non-uniform distribution of plating thickness causes an extra burden on the subsequent CMP process. Thus, it is desirable to make the plating surface as planar as possible.

Various processes for planar plating

Various types of processes for planar plating have been developed, but any of them has not actually been adopted. It is because, regardless of plating methods, such as use of different chemicals or scrubbing, the quality of deposited film is adversely affected (this conclusion is based on the results of my own experiments, not considering other companies' systems).

Therefore, another idea to facilitate planar plating without chemical use or scrubbing has come up.

Porcupine process

The figure on the next page shows the steps of the porcupine process. In this process, a porous pad is placed on the wafer surface for plating. The plating grows much in the through-holes of the pad and in the wafer trench (plating target), but less in the other area.

Cu posts formed by the through-holes of the pad are like spines of a porcupine and can be easily removed by etching. Etching must be performed without damaging Cu film in the interconnect trench. As a result, the Cu film left in the area other than the trench is very thin and planar.

This technology is in a development stage. If successfully developed, it may be an innovative technology for the 32 nm generation or beyond that will reduce the burden on the CMP process and dramatically improve planarization performance.

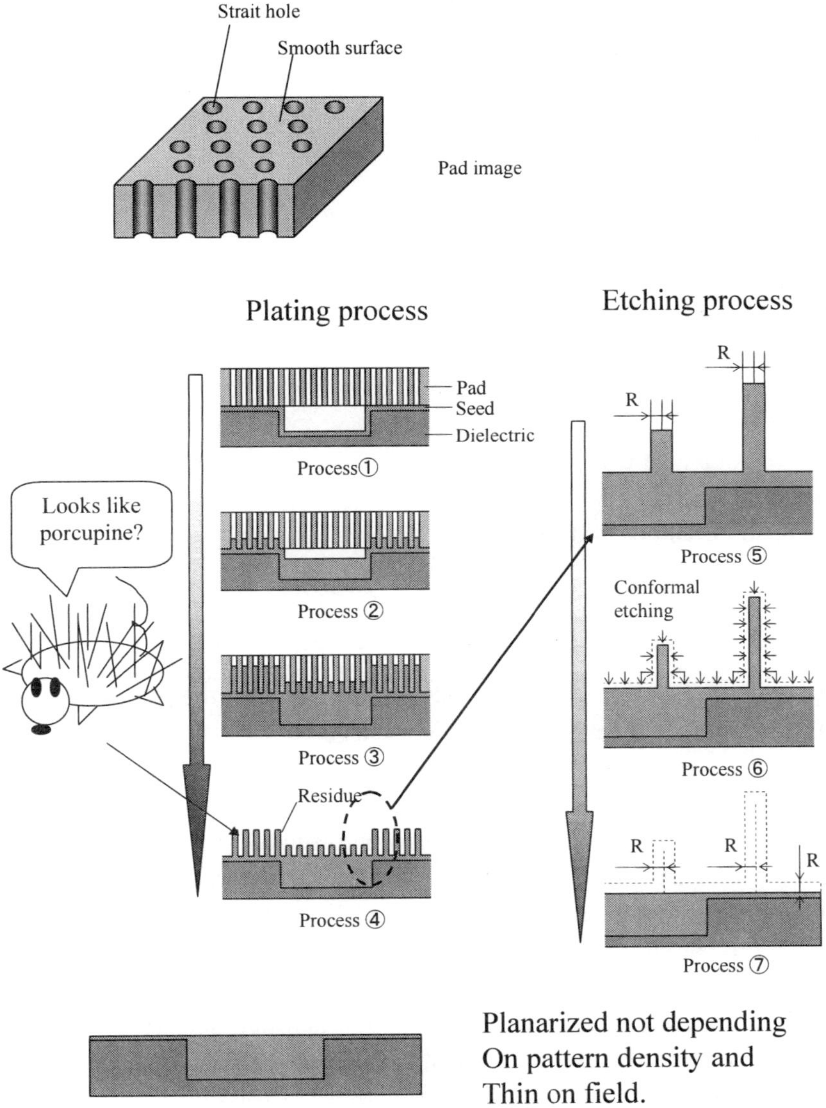

Planarized not depending
On pattern density and
Thin on field.

Metal cap plating

Specifications and use of a cap metal

The specifications and use of a cap metal layer are as follows.

(1) Low resistance is an essential requirement for being a part of interconnects.

(2) To reduce the effective dielectric constant, deposition should be made in the interconnect region only. In other words, it is desirable to use no inorganic cap, including Si-N (this requirement has not been achieved yet).

(3) A cap metal prevents electro migration (EM) and stress migration (SM). Since a Ta-N barrier layer inherently adheres to dielectric film, low adhesion between the barrier layer and Cu may cause EM or SM. Adhesion can be enhanced by a cap metal to improve EM and SM resistance.

(4) Enhanced adhesion between the barrier layer and Cu reduces leakage through this surface.

(5) A cap metal provides protection upon dielectric film etching.

Various types of caps made of W, Co-W, Co-B, etc. are currently available.

Co-based cap formed by electroless plating

There have recently been many papers dealing with the improvement of EM resistance. The improvement of EM resistance with the use of Cu alloy or future carbon nanotubes (CNT) has been suggested. It has also been reported that, with Co-W cap film, EM resistance is increased by 100 times.

The figure on the next page shows an example of Co deposition linearly controlled with time in a range between 5 and 20 nm (in actual operation, Co deposition is made in a range between 5 and 10 nm). For electroless plating, this control level can be achieved as long as the temperature and flow in the bath are precisely controlled.

The key for electroless plating technology lies in selectivity. It is desired to form cap metal film only over Cu interconnects; any cap metal film left on dielectric film causes deterioration in reliability. Basically, electroless plating is suitable for non-selective deposition. Each manufacturer has its own know-how to secure selectivity in electroless plating. SEM photos on the next page show selective Co deposition on Cu interconnects and dielectric film.

Several types of metal cap film, such as Co-W and Co-B, are available depending on the type of plating solution.

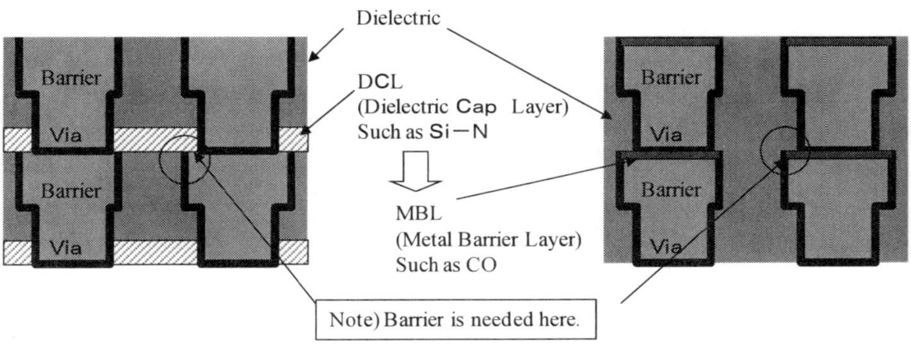

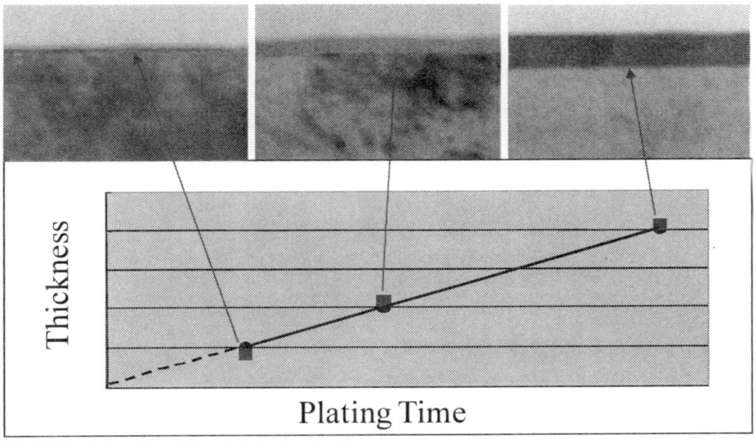

Thickness control

SEM

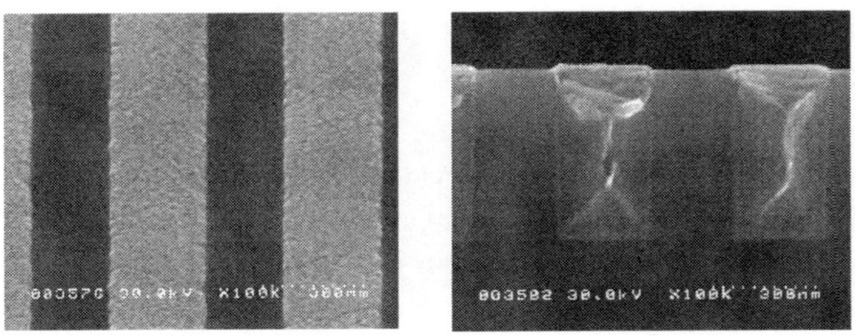

Bump plating

Bump plating

Bumps are electrodes formed on the uppermost layer of devices for external connection. The figure on the next page shows a cross sectional view of a bump.

The bump process flow is shown in the figure. The bump position is first defined by a resist, and plating is performed to fill the resist opening. Then, resist removal and reflowing are conducted to form the bump, as shown in the figure. Often being made of solder or gold, bumps are formed by electroplating.

The bump size varies between 50 and 100 μm square depending on the device generation. There is a roadmap that proposes to reduce the bump size to about 10 μm square.

Many competing technologies, such as printing or mounting ball-shaped bumps on specified holes, are available. Competition among such technologies in terms of throughput, precision, and cost is tough. To meet demands for scaling, a method suitable for microhole filling will be adopted in the future.

Image of bump plating systems

The figure shows a schematic diagram of a multi-bath electroplating system. Interconnect plating only requires a Cu plating bath. Bump plating, which is closer to device implementation, requires (1) surface treatment before plating, (2) rinse, (3) plating (deposition of several different films), and (4) cleaning. Since plating takes long time, a multi-bath system is generally used for efficient plating. Thus, bump plating systems tend to require a larger installation space.

Plating system manufacturer and plating service provider

Who provides bump plating? While some device manufacturers perform bump plating themselves, there are many plating service providers called *bump houses* who provide bumping services.

Plating is not a unique technique in device implementation. Bump plating in the wafer process is different from implementation, and it is not as complicated as multilayer interconnection. Bump plating is a process placed between multilayer interconnection and implementation. In fact, device manufacturers seem to be uncertain about whether 10 μm bump plating in the future should be handled on the implementation side or the multilayer interconnection side.

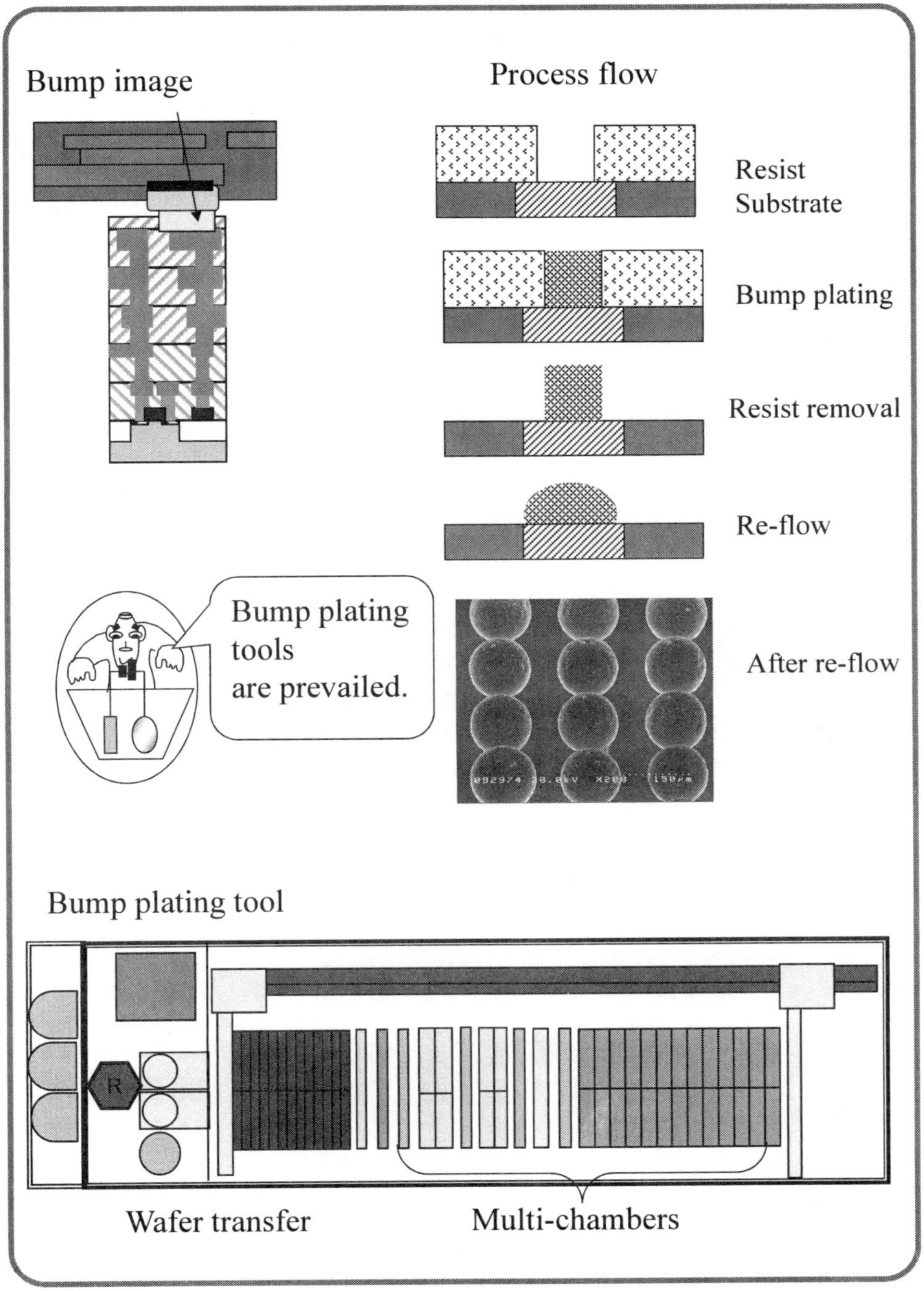
Bump image
Process flow
Resist
Substrate
Bump plating
Resist removal
Re-flow
After re-flow
Bump plating
tools
are prevailed.
Bump plating tool
R
Wafer transfer
Multi-chambers

= Manabu-style: Power Pointist =

Soon after I joined Ebara Corporation in 1974, I was appointed to be in charge of construction of a hydraulic power plant in Indonesia. English was the language used between our Indonesian counterparts and us for communication. At that time, I could only speak beginner-level English, such as "Me, engineer. You, customer. Understand?" It was tragic and comedic at the same time, since our counterparts were no better than me. I thought that it was a miracle that we could construct a power plant under such incredible circumstances. Recently, however, I have changed my view. I began to feel that speaking poor English was tremendously fortunate for an engineer. Why is that? First of all, telling a lie in English is impossible. Secondly, when you try to convey what you wish to express, you will stress that expression repeatedly until your listener understands. I learned to show what I wished to express as an engineer by drawings and to extract essential elements through such experience.

How have I developed semiconductor process systems? I drew what I was trying to express. It helped me understand the essential elements of development. It worked like magic. Since then, I have always started with punch drawings. I made it a custom to draw illustrations for all of my writing. I do not mean to be too personal, but my sister, Hiroko Tsujimura, is a *famous* Manga artist (cartoonist). In rivalry with my sister, I created my style for illustrations by making full use of PowerPoint. I named this style *Power Pointist*. Anyone who wishes to be my student will be able to draw Manga soon.

Indonesian engineer   Manabu in young age   Manabu now!

Chapter III

Wet Removal Technology : CMP

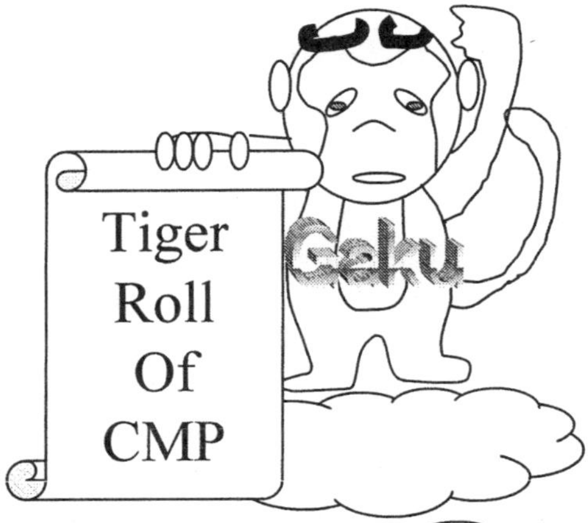

Stone to Devices

From the world oldest polishing to the latest polishing (CMP)

In the previous chapter, plating is mentioned as one of the oldest technologies in the world. If plating is the black king on the chessboard, polishing is the white king. The origin of polishing technology goes back to polished stone tools, which were already available in the Jomon period in Japan. Basically, polishing a substance is performed by using the same substance. For example, SiO_2 is polished by using SiO_2. In addition to SiO_2, tungsten and Cu are polished for semiconductor devices. However, tungsten is not polished by tungsten, and Cu is not polished by Cu. In this respect, CMP seems to go beyond the border of polishing. Nevertheless, polishing is still a major technology in semiconductor device planarization, and further requirements for the polishing of diverse materials will emerge in the future. Applying polishing technologies developed in the semiconductor field to the conventional polishing field may also be possible. The world of polishing is filled with hopes.

Overview of the history of CMP technology

CMP stands for Chemical Mechanical Polisher and is an indispensable system for today's planarization process for semiconductor devices. It was originally developed and adopted by IBM as a *planarization process for logic devices* in the late 1980s. Compared to memory devices, logic devices have a larger number of layers, and planarization requirements for them are stronger. For logic devices, the chip production quantity is also smaller with a higher chip cost. Therefore, it seems that CMP for offline use (outside clean rooms) was easily employed for logic devices. Subsequently, Intel, Motorola, AMD, and other companies in the U.S. began to adopt CMP. Japanese device manufacturers at that time did not employ CMP since their main products were memory devices, and planarization requirements were satisfied with other planarization technologies, such as spin-on glass (SOG: method to form dielectric film by coating) or etch-back. In those days, CMP was based on bare silicon wafer polishers and was not designed for installation in clean rooms. Considering that a special room for CMP was required, Japanese device manufacturers hesitated to fully introduce CMP systems.

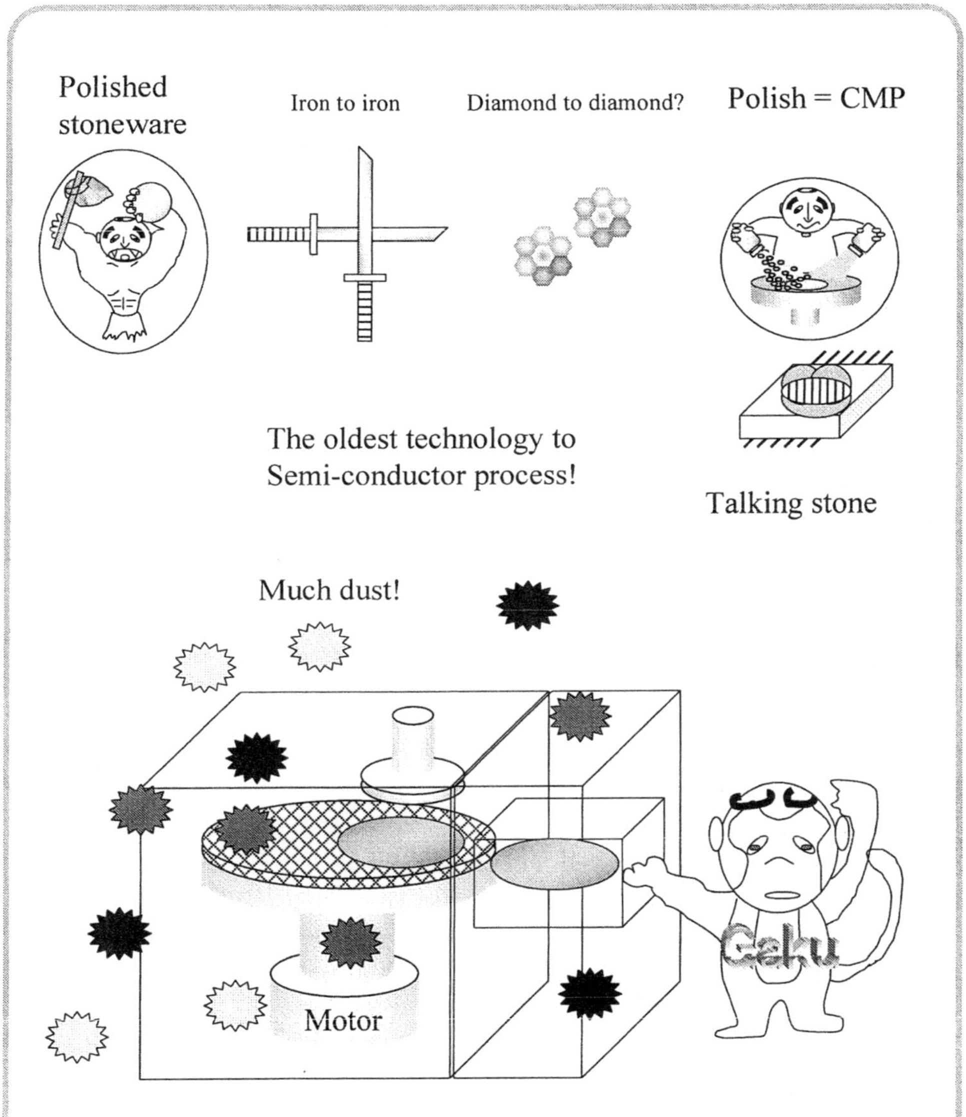

Dirty, Hard and Experimental : Three demerits

Dry-in/dry-out comes on!

Other planarization technologies

Major planarization technologies competing with CMP in the early days are as follows.

[ILD planarization]

(1) Bias CVD method that performs deposition with a RF bias applied to wafers

(2) Thermal CVD method with $TEOS/O_3$ flow

(3) SOG method by spin-coating

[Interconnect planarization]

(1) Heating and fluidization after Al deposition

(2) Etch-back after blanket CVD of W

These technologies were applied to 350 or 250 nm technical nodes in the 1980s and the early 1990s. However, they have been replaced by CMP since the emergence of the damascene process.

Dry-in/dry-out

As scaling requirements for semiconductors became more stringent in the early 1990s, CMP technology came to the fore in Japan. During this period, there was a significant innovation in the design concept of CMP. That is the dry-in/dry-out concept.

After it was demonstrated that CMP based on this concept could be used in clean rooms like dry process systems, the use of CMP accelerated. CMP became a common technology mainly for ILD planarization in the generation of 200 mm (wafer diameter)/250 nm (technical node). Then, we entered the new 300 nm/130 nm generation.

The major features of this new generation are the full introduction of the *dry-in/dry-out* concept for 300 mm wafers in the mechanical aspect and the adoption of diverse processes (STI, damascene Cu interconnect, low-k material, etc.) for the 130 nm technical node in the process aspect. The era of regarding CMP as special (polishers requiring less cleanness) will end, and requirements for CMP will be more stringent than those for dry processes in terms of process and mechanical performance. The ITRS, a global standard for the development of semiconductor devices/systems, has stated that CMP is an indispensable technology for the next generation devices. CMP has begun to receive global recognition.

Other planarization technologies

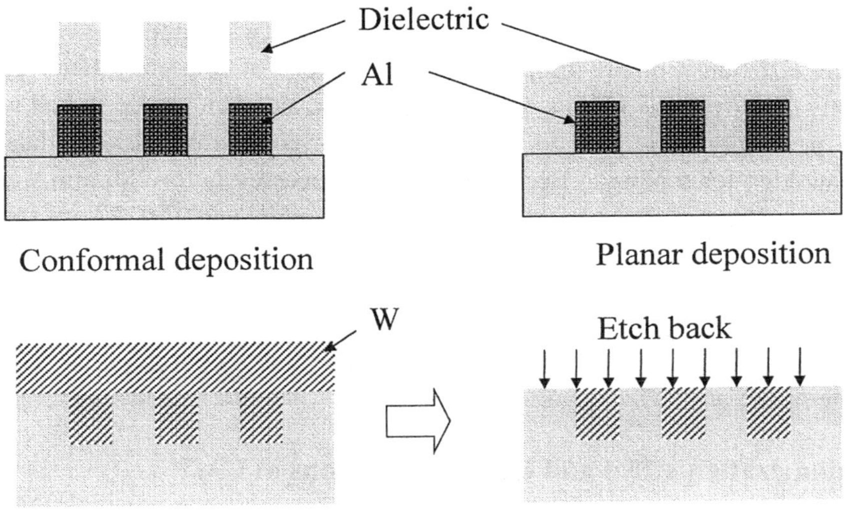

Dry-in and Dry-out technology

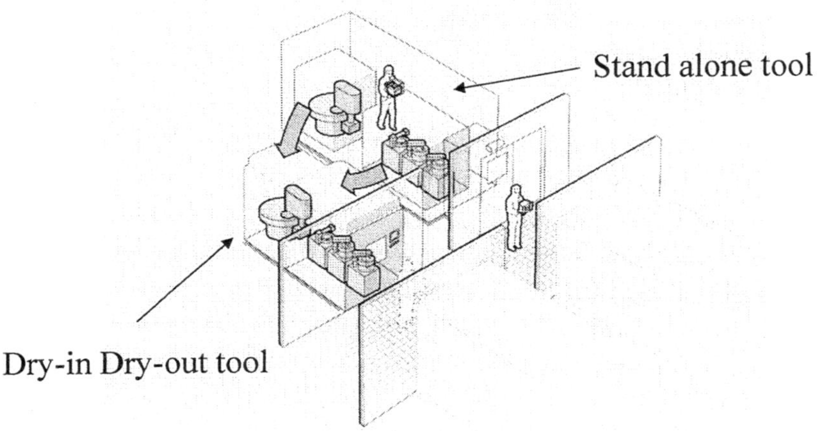

CMP like a drug

CMP to meet requirements for scaling and wafer size transition

The figure on the next page shows how CMP was adopted as scaling and wafer size transition proceeded. In the generation of 200 mm wafers, CMP was partly employed for 350 nm generation devices, and then its use spread with 250 nm generation devices. Since then, CMP is considered as an indispensable technology for scaling. Because the same processes as for 200 mm wafers are used for 300 mm wafers, CMP is indispensably employed for 130 nm generation devices.

At present, CMP is used in various processes, including ILD planarization, tungsten contact (via), STI, polysilicon planarization, and Cu damascene, for the 45 nm generation or beyond. It has become an indispensable technology for state-of-the-art device manufacturing.

Planarization effect and other applications of CMP

The figure compares planarity between two devices; one is processed with CMP, and the other is processed without CMP. For multilayer interconnects the planarization of each layer is increasingly important. CMP is also useful for surface modification. Various applications of CMP may be possible, such as removing nodules left on each layer by polishing or enhancing the cleaning effect by polishing only with D.I. water.

CMP is like a drug!

The term *drug* seems inappropriate to describe CMP, but the phrase above is the precious words of an engineer actually using CMP. It was a story in the early 1990s when people stayed away from CMP, saying "CMP? That dirty process?" In those days, no one thought of CMP becoming widely used. At first, the engineer tried CMP with hesitation. Once using CMP, however, he changed his view of CMP. He found that, although the device surface was contaminated, the contamination did not reach the inside of devices and that planarization performance was *excellent*. One year later, he became a top engineer in the CMP field.

Recently, he declared that he could not fabricate any device without CMP. At the same time, he said that CMP was like a drug, as presented in the title of this section. His words remain and will forever remain in my diary.

Scaling and Bigger wafer trends and CMP application

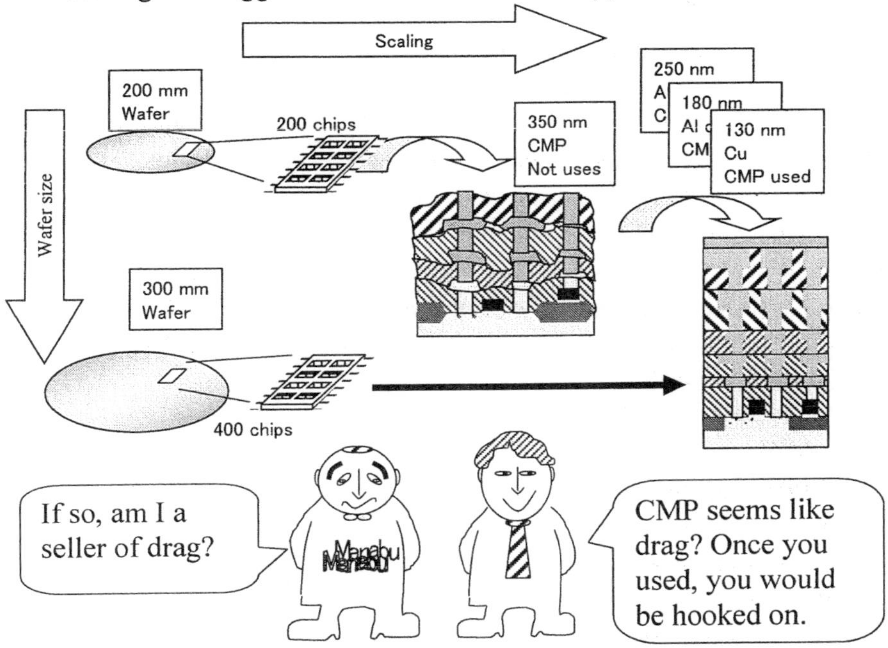

Planarization image by CMP

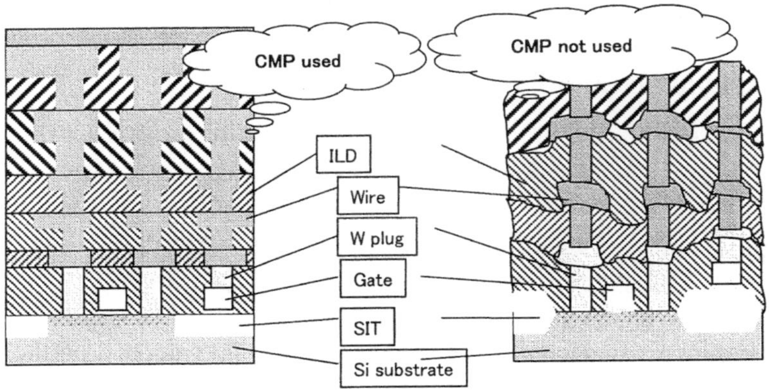

Let's start with Preston's Law

Preston's equation

The principle of CMP is explained below. CMP is an application of the conventional polishing process used for bare silicon wafers. The explanation below is given with commonly used rotary CMP taken an example. A wafer is held by the wafer carrier with its top surface facing downward. Slurry (containing a few % of abrasives, such as silica) is dispersed onto the pad on the rotary table. The wafer is sandwiched between the backing film and the pad. Then, it is polished by being pressed against the pad by the wafer carrier. During this process, both the rotary table and the wafer carrier rotate in the same direction at almost the same speed. The polishing rate is proportional to the load and relative speed.

Polishing follows Preston's equation as follows.

$PR = kPV$

PR: Polishing rate (m/s)

k: Constant (Pa^{-1})

P: Polishing pressure (Pa)

V: Relative speed (m/s)

Stable polishing pressure is ensured by the head design. The same polishing rate can be obtained anywhere on the wafer if both the rotary table and the wafer rotate in the same direction at the same speed.

PVT dependence and PVTQ dependence

If chemically active slurry is used, the polishing rate is also affected by temperatures. In such cases, the following modified equation is used.

$PR = kPVT^{\alpha}$

T: Temperature during polishing (°C)

α: Temperature index

There is a recent trend of decreasing slurry flow to reduce CoC (cost of consumables). In such cases, the polishing rate may depend on the slurry flow, and the following modified equation is used.

$PR = kPVT^{\alpha}Q^{\beta}$

Q: Slurry flow (m^3/s)

β: Flow index

As described above, the original Preston's equation and its applied equations are practically used.

CMP Principle

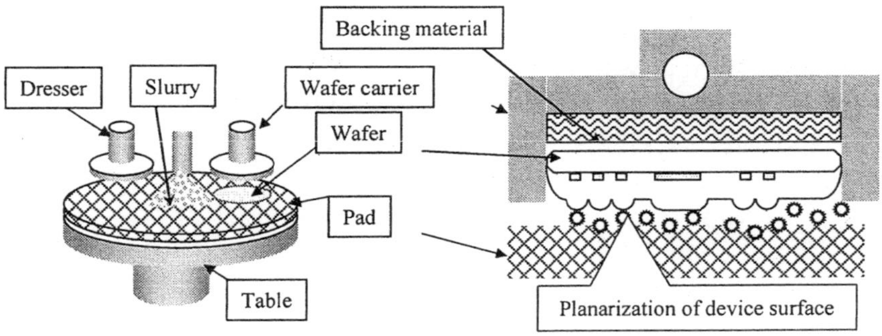

Dr. Preston
(1896 – 1989)

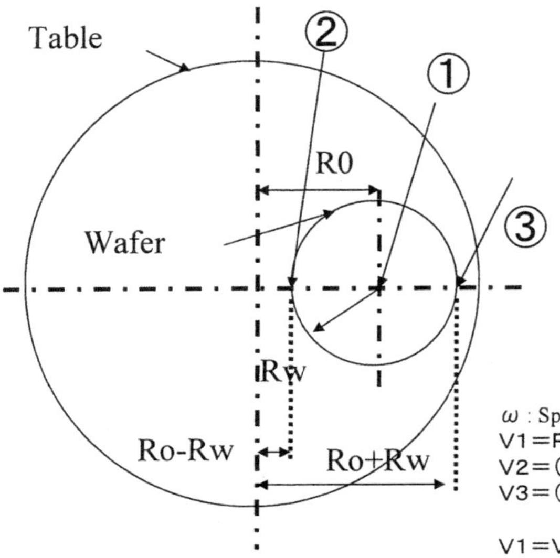

ω : Speed of table and carrier

$V1 = R0 * \omega$

$V2 = (R0 - Rw) x \omega + Rw x \omega = R0 * \omega$

$V3 = (R0 + Rw) x \omega - Rw * \omega = R0 * \omega$

$V1 = V2 = V3$

This is CMP!

Simple explanation of the planarization principle

This section explains what CMP is in a simple manner. More precisely, CMP is not so simple, but a simplified explanation is easier to understand for beginners of CMP. CMP is a polishing process combining chemical and mechanical factors. Let's learn what chemical factors are and what mechanical factors are.

(1) Oxide film

Basically, SiO_2 film (Si di-oxide film) is polished by using abrasives made of the same material (SiO_2). Polishing with the same material may reduce scratches. To further reduce scratches, alkaline solution, which dissolves SiO_2, is added to smooth the surface. The use of abrasives is a mechanical factor while the addition of alkaline solution for surface smoothing is a chemical factor.

SiO_2 (SiO_2 abrasive + alkaline solution) -> $SiO_2\uparrow$

(2) Metal (tungsten)

The polishing of a pure metal (W, etc.) with abrasives causes scratches (due to metal softness or viscosity). To prevent scratches, the W surface is oxidized and then mechanically polished as in the case of SiO_2 polishing. This oxidization process is chemical while the polishing process is mechanical.

W + oxidizer -> $WO_X\uparrow$

(3) Cu

Basically, the process is the same as for W. However, Cu oxide film is not as strong as W oxide film. Just imagine that the film is easily corroded. Instead of forming oxide film, Cu is mechanically polished after complexation (oxidized at the same time). This action is chemical while the polishing of formed Cu complexes is mechanical.

Cu + oxidizer -> Cu^{++} (Ionization, instead of forming CuO_X, is also regarded as oxidization.)

Cu^{++} + complexing agent -> Cu (complex)$\uparrow$

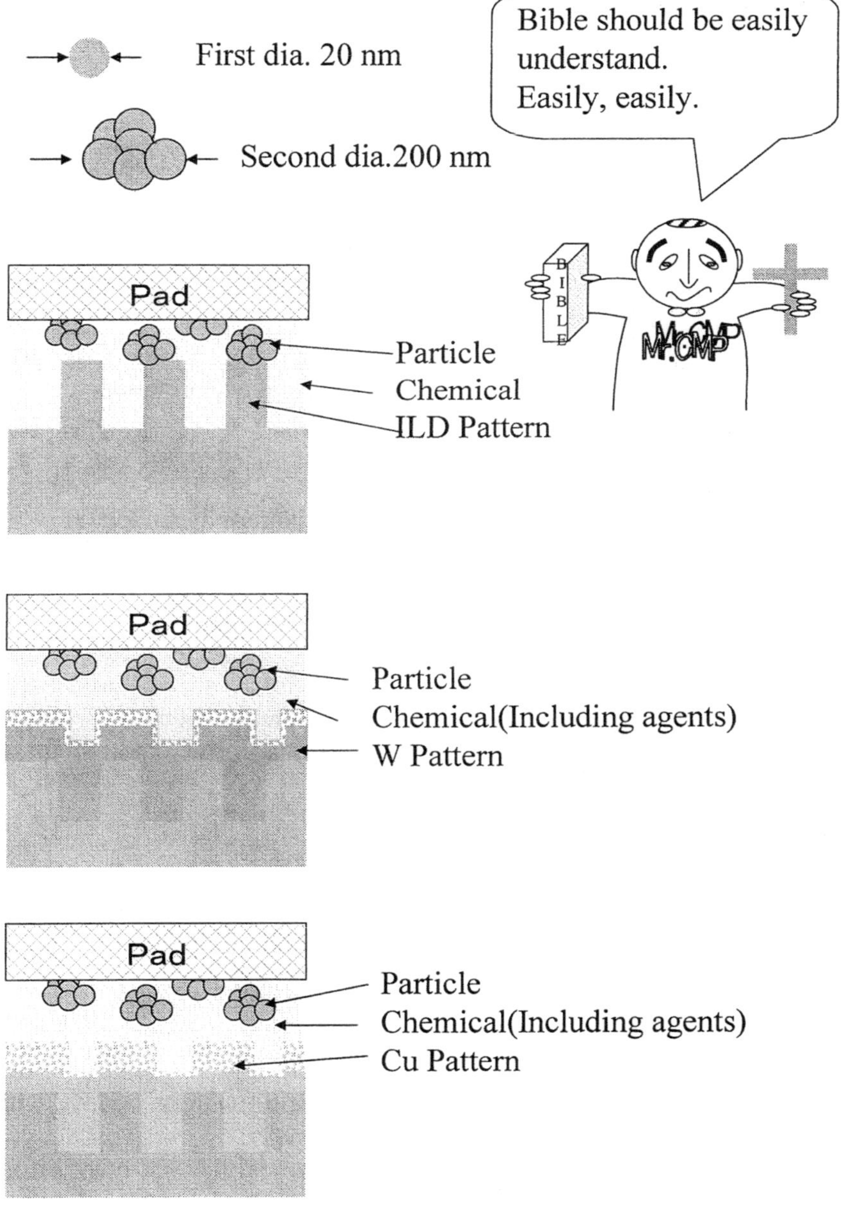
Bible should be easily understand.
Easily, easily.
First dia. 20 nm
Second dia.200 nm
Pad
Particle
Chemical
ILD Pattern
Pad
Particle
Chemical(Including agents)
W Pattern
Pad
Particle
Chemical(Including agents)
Cu Pattern

Do you oppose Preston's Law?

Preston region and non-Preston region

Polishing controlled by Preston's equation is called as polishing in the Preston region. Recently, polishing beyond this theoretical equation is observed. It is called as polishing in the non-Preston region and is mainly used with chemically active slurry. In short, it has high etching performance.

Polishing is affected by chemical and mechanical factors. Which factor is more dominant? How is chemical reaction/mechanical processing made uniform across a wafer/chip? They are the themes of CMP technology.

Simplifying performance parameters

The performance of semiconductor manufacturing systems is generally categorized into process performance and mechanical performance.

Process performance parameters

- Processing rate
- Wafer level non-uniformity
- In-chip non-uniformity (deposition/polishing pattern dependence)
- Defects (FM, scratch, etc.)

Mechanical performance parameters (for the process above)

- Process capability: throughput (wph) and tact time (min)
- Reliability
- Productivity index

CMP performance is evaluated by these parameters. Let's study CMP based on the common knowledge above with some CMP-specific requirements added.

Evaluation items

(1) Polishing rate: The higher, the better. The polishing rate per minute is used for reference. If the film thickness to be polished per minute is 1 μm, the polishing rate is 1 μm/min. A general guideline is as follows.

(1) ILD > 500 nm/min (1000 nm/min in some cases)
(2) STI > 200 nm/min
(3) W > 500 nm/min
(4) Cu > 500 nm/min

(2) In-wafer non-uniformity - Verified with solid wafers: $3\sigma 3$ % with 3 (2) mm EE

Oxide film CMP is greatly affected by mechanical factors; non-uniformity is directly susceptible to head performance. Metal grinding is greatly affected by chemical factors; the critical point is how uniform chemical reaction is achieved.

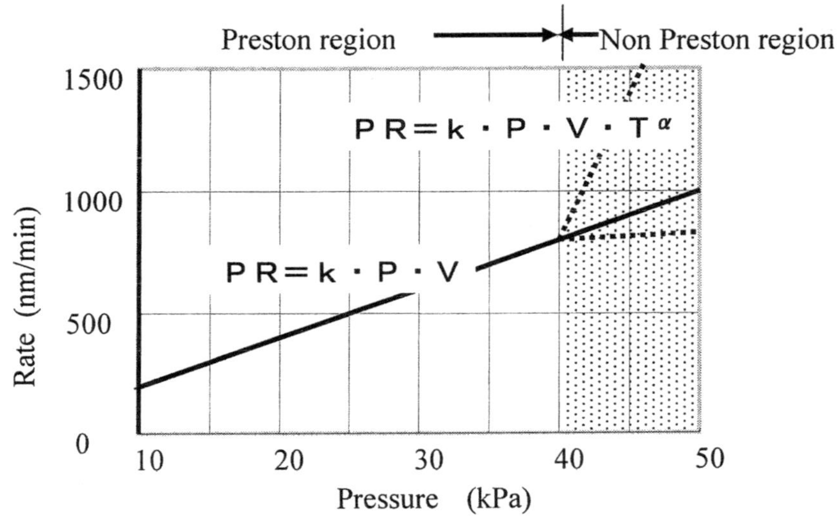

PV range of several processes

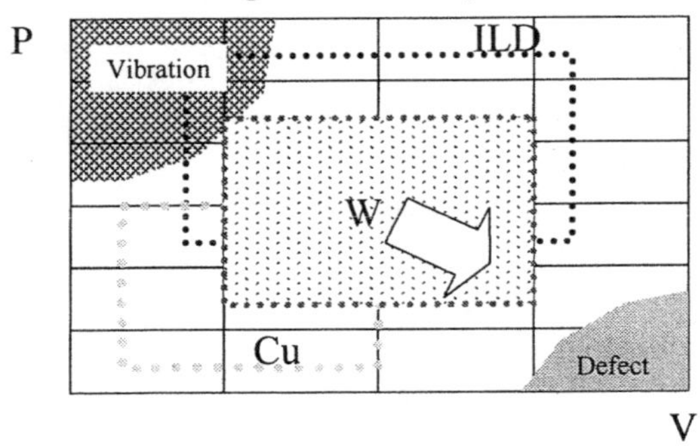

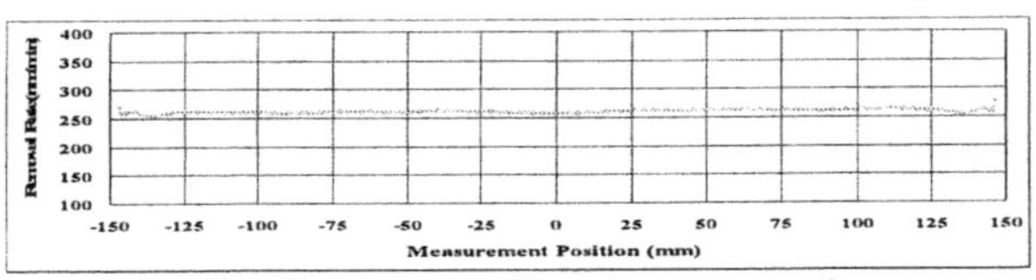

Wafer non-uniformity example

I will give you the secret of CMP.

Polishing peculiarities?

Attention! Uniform polishing cannot be achieved with evenly applied load!

Reason 1: Polishing produces a variety of basic processing profiles depending on the wafer pattern profile of processed articles and the slurry/pad type. The variation of basic processing profiles is called *polishing peculiarity*. It is the ultimate secret of polishing to produce a constant polishing profile by controlling (changing) the polishing pressure to correct polishing peculiarities.

Reason 2: Various profiles of incoming wafers are used, and demands for controlling such profiles have been increased. It is called profile control in polishing. As mentioned in Reason 1, polishing basically involves profile control. Therefore, the process can adjust the profiles of incoming wafers even if they vary a little.

Generally, current CMP systems achieve desired performance by controlling the polishing pressure to a preferred level depending on the device pattern and the slurry/pad type.

Other variable factors

Note that polishing peculiarities include many other variable factors that affect polishing, as outlined below.

For example, the dresser type, dressing method, and device film quality greatly affect the polishing rate and polishing profile. Years ago, the distribution of B and P on BPSG film and Cu annealing temperature were also included in such factors. In recent years, attention must be paid to etching damage to low-k material as an affecting factor.

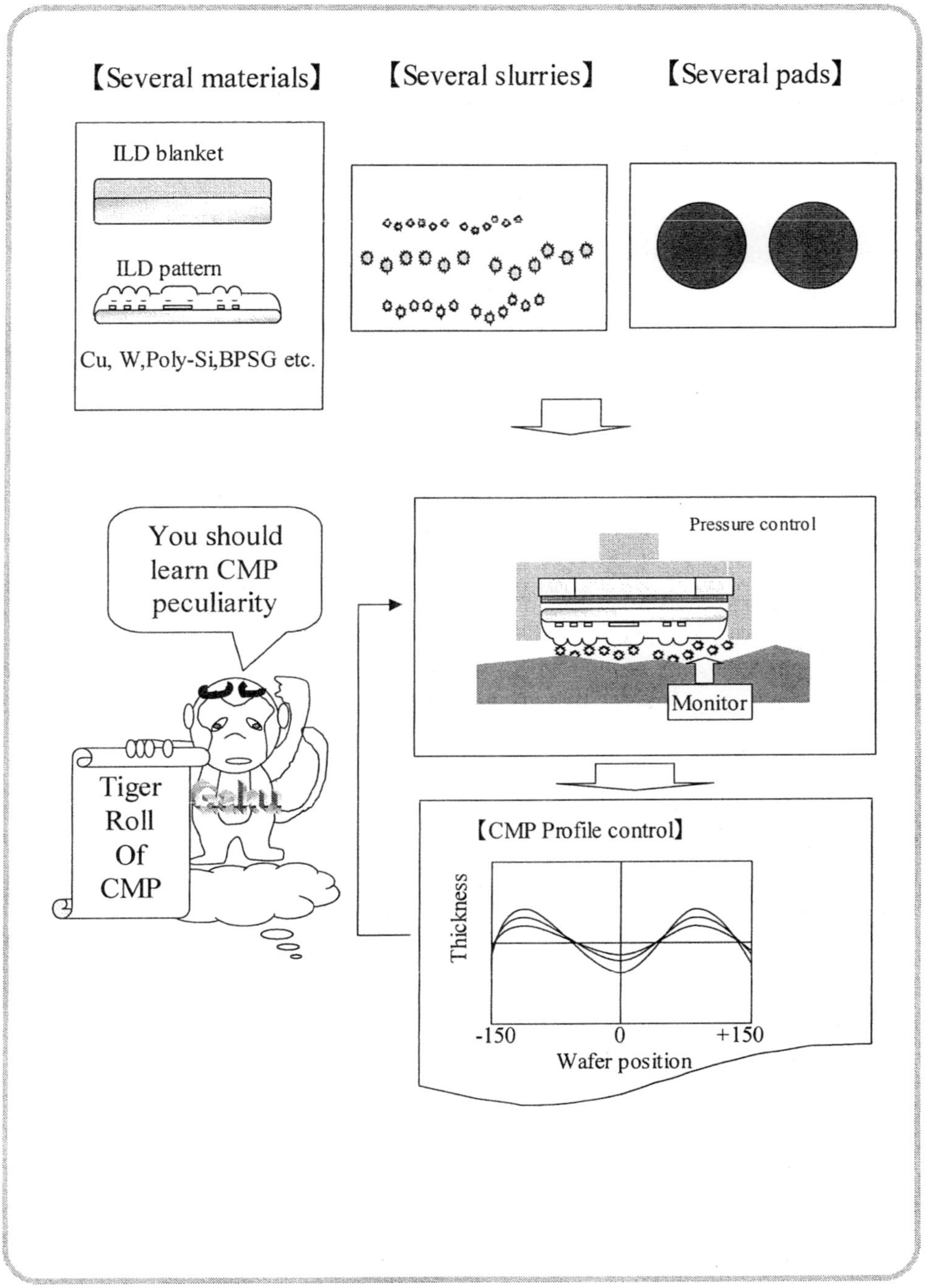
【Several materials】
【Several slurries】
【Several pads】
ILD blanket
ILD pattern
Cu, W,Poly-Si,BPSG etc.
You should learn CMP peculiarity
Tiger Roll Of CMP
Pressure control
Monitor
【CMP Profile control】
Thickness
-150
0
+150
Wafer position

Eternal problem: pattern dependence

Planarization performance (dishing, erosion, etc.)

According to the interconnect pattern, a chip may contain some regions with a high density of concave/convex parts and other regions with a low density of them. This fact means that, regardless of the deposition method, deposition is associated with dependence on the pattern. It is called the pattern dependence of deposition. In the same manner, the removal of resultant profiles is dependent on the pattern. It is called the pattern dependence of deplating.

The plating of wide, dense, or sparse patterns, as shown in the figure on the next page, characteristically results in (1) dishing caused by the plating process and (2) mounding on the region of dense patterns.

For such patterns, CMP results in (1) dishing caused by the CMP process in wide patterns, (2) erosion of oxide film, which easily occurs in dense patterns, and (3) recess (overpolishing) of the entire area.

Note) Deposition, removing, and surface treatment suffer from the same problem, pattern dependence. Devices have various types of patterns, and pattern dependence in related processes cannot be solved in any way. It is desired to develop processes that are less affected by pattern dependence.

Defects - Categorized into FM, PS (scratch), PR (residue), etc.

The key for defect inspection is what type of defects is measured and what type of inspection systems is used for the measurement. Currently, inspection systems are categorized into three major groups.

1) FM inspection system for blanket film

2) Pattern defect inspection system -> Pattern defects are categorized as above into FM, PS, erosion, etc. Bright-field and dark-field types are available, and both are required in principle.

Generally, defects are categorized into:

- FM (Foreign materials)
- PS (Polish scratches)
- Polishing residue
- Corrosion
- Erosion.

Each inspection system supplier gives them unique names.

Planarization performances of plating and CMP

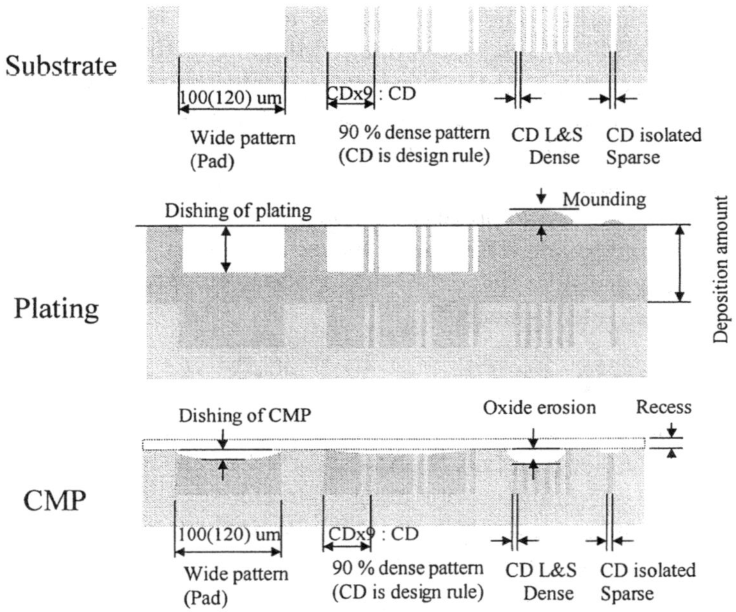

Defect classification

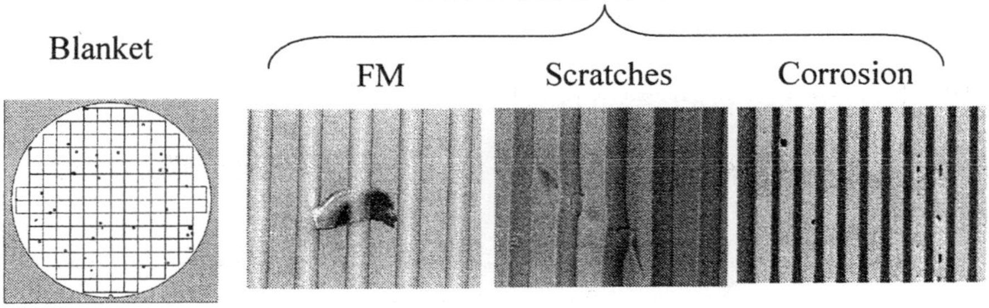

Where in devices?

Applicability of CMP

The figure on the next page shows a device containing STI, ILD, tungsten (W) plugs, and metal layers. The CMP process for theses layers can be roughly divided into *a blind CMP process* without CMP stopper film and *a recess CMP process* with stopper film. ILD CMP is a blind CMP process for homogeneous ILD, which is primarily required for step height reduction. STI CMP is a *recess* CMP process with stopper film, such as Si-N film, which is required to improve the uniformity and reduce the dishing of stopper film. Interconnect polishing is a *recess* CM P process, including barriers, which is required to reduce dishing and oxide erosion. Dishing refers to the overpolishing of interconnects to a dish-like shape. Oxide erosion refers to the overpolishing of oxide film between interconnects.

Four stages of CMP

As describe above, CMP is used for oxide film (ILD and STI), metal film (W, AL, and Cu), and other materials (PolySi, etc.). For generalization, this section categorizes CMP according to polishing characteristics and materials. Blind CMP is for ILD while *recess* CMP is for STI, W, and Cu.

CMP-1: Step height reduction (ILD) (homogenous)
= Capability of step height reduction
CMP-2: Bulk polishing (homogenous) = Wafer non-uniformity
CMP-3: Inhomogeneous polishing (recess stopper/barrier) = Selectivity
CMP-4: Underlying material polishing = Defect elimination

The stage CMP-0 also exists in actual polishing. The first problem is caused by polishing the oxide film of the processed material. For Cu film, CuO_X polishing is required prior to Cu polishing.

Difference between CMP processes for bare silicon wafers and device wafers

For reference, a major difference between CMP processes for bare silicon wafers and device wafers is explained. The former polishes waves on both sides of wafers by about 20 μm to make the wafer thickness uniform and to reduce surface roughness. The material to be polished is silicon only. The latter, as described above, polishes one side of wafers by about 2 μm to keep a film thickness accuracy of 20 nm and to polish various inhomogeneous materials. Thus, they are considerably different from each other.

CMP application

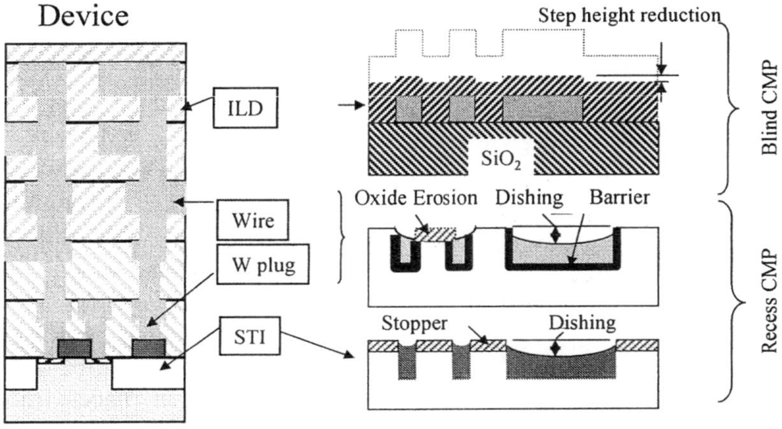

Several CMP processes

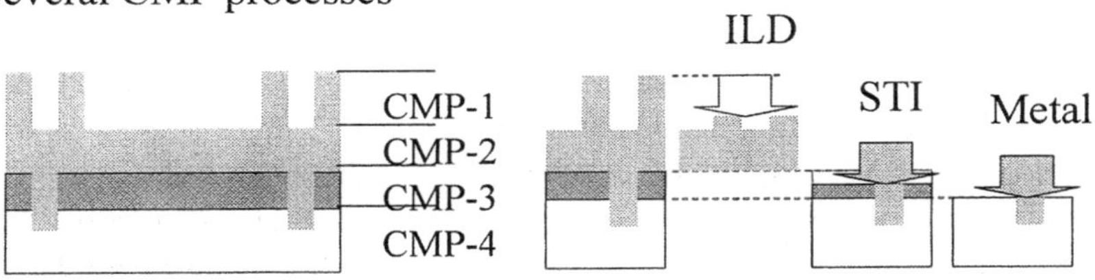

Difference between bare-Si polish and device polish

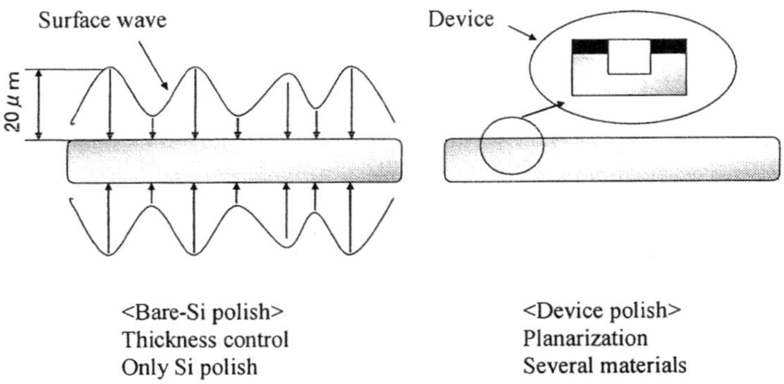

Polishing and cleaning subsystems

A CMP system is comprised of polishing, cleaning, drying, monitor, slurry supply, waste treatment, pad, slurry, control, and automation technologies. This section particularly explains the main components: polishing, cleaning, and monitor subsystems.

Polishing subsystem

In the 1980s, polishing technology started with bare silicon wafer polishing; rotary type polishing equipment was used in most cases. Throughput was more weighted than performance, and new types equipped with multiple heads or tables were developed one after another. In the 1990s, as scaling and higher planarization levels were required, single-head type systems were preferred because of their ease of control. Also, every kind of prototype polishing systems emerged during this period.

(1) Small table type: Developed for space saving. It places wafers over the center of the rotary table and polishes them. Geometrically, the on-wafer relative speed can be constant even if wafers are placed over the table center.
(2) Small head type: Places wafers with their top surface facing upward and polishes them using a head smaller than them.
(3) Linear type: Moves the pad in a linear (not rotary) motion.
(4) Grinding type: Performs planarization by grinding (not by polishing).
(5) Fixed abrasive type: Polishes wafers with fixed abrasives (not loose ones).

Cleaning subsystem

In the 1980s or even in the 1990s, there was no idea of incorporating built-in cleaning subsystems into CMP systems. Polishing generates enormous dust. On the other hand, a cleaning subsystem must meet the severest cleanness requirements among all subsystems for semiconductor manufacturing. It was often said that combining polishing and cleaning subsystems was like bringing the sky and the ground together. Still now, no chemical engineering process can offer a particle separation process that reduces the particle count of 10^{15} to the order of 10^{9}. Thus, conventional batch process type RCA cleaning was counted on. As the application of CMP increased, people naturally started to want to use CMP in the same manner as for dry process systems. In the 1990s, a concept called *dry-in/dry-out* emerged. Based on it, vertical type or one-bath type cleaning subsystems were developed. Functional water for processing without chemicals has also emerged, such as ozonized and ionized water.

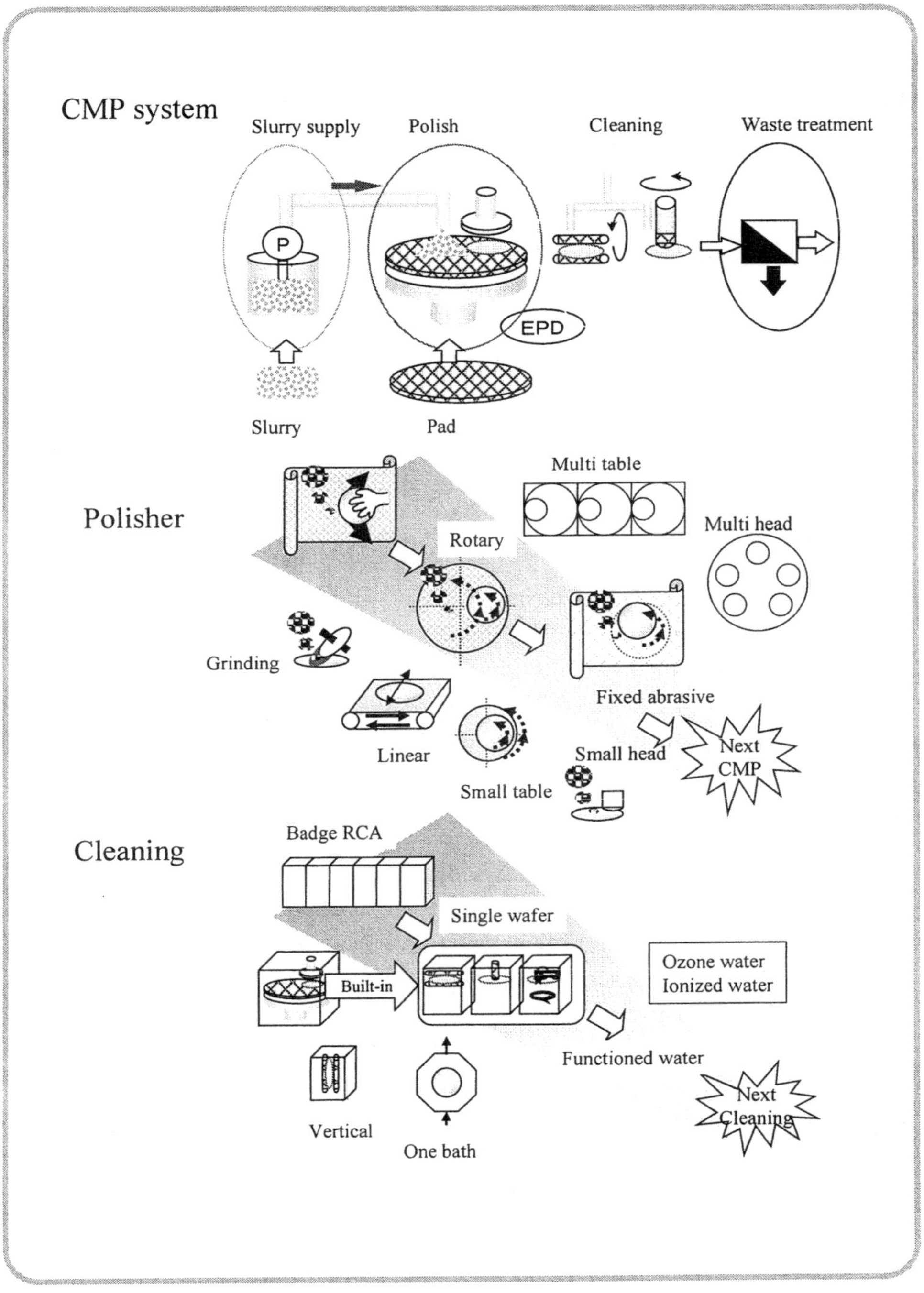
CMP system
Slurry supply
Polish
Cleaning
Waste treatment
P
EPD
Slurry
Pad
Polisher
Multi table
Rotary
Multi head
Grinding
Fixed abrasive
Linear
Small head
Next CMP
Small table
Cleaning
Badge RCA
Single wafer
Ozone water
Ionized water
Built-in
Functioned water
Vertical
One bath
Next Cleaning

Monitor subsystem

Monitor type

Polishing is roughly divided into the blind CMP process for polishing homogenous film (ILD, etc.) and the recess CMP process for polishing inhomogeneous film (W, Cu, STI, etc.). Various monitor methods have been developed to support these processes.

(1) Friction detection method: Detects torque variations at inhomogeneous film boundaries with a torque detector set to the table or head motors. It is often used for the W process. The motor current method is in this category.

(2) Vibration detection method: Detects vibration acceleration variations at inhomogeneous film boundaries with a vibration acceleration monitor set to the head.

(3) Eddy current detection method: Monitors film thickness variations by using eddy current that varies according to the metal film thickness. It is used for the recess CMP of metal film.

(4) Optical detection method: Based on the same mechanism as that of the optical film thickness monitor. In-line and in-situ types are available.

(5) Others: Various methods have been developed, such as detecting special chemicals generated upon inhomogeneous film polishing or detecting acoustic variations.

Comparison of actual data

The figure on the next page shows data obtained by the methods of motor current, vibration, and eddy current, which are applied in actual manufacturing. The motor current method is the oldest and detects frictional force between inhomogeneous films based on a rotational torque. The vibration method detects the difference in the vibration pattern between inhomogeneous films with a vibration sensor on the head. The figure compares the results of device pattern detection between the motor current method and the vibration method. The eddy current method is based on a principle that the level of eddy current generated in metal film varies according to the metal film thickness. The figure shows an example of Cu and Ta polishing.

Monitor

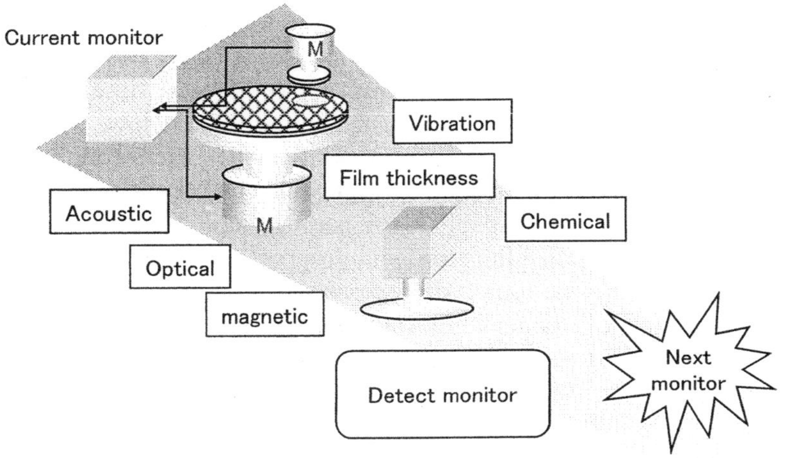

Comparisons

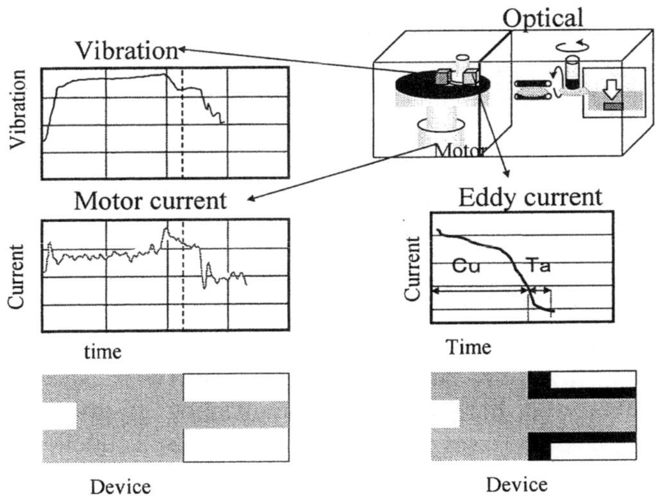

Raw process time

Improvement of raw process time

In-process monitoring mainly aims to monitor polishing stop information and film thickness control information. It is also effective for failure diagnosis, lifetime prediction, raw process time (real process time) reduction, etc. In-process monitoring for polishing systems is outlined below.

1) Polishing stop information (end point): CMP offers rough processing as described above and requires end point control techniques. Mechanical stop methods are available, as illustrated herein. Slurry/pad development for automatic polishing stop is in progress. Preferable end point control techniques for the four CMP regions shown in the figure are required.
2) Film thickness control information: ILD film thickness greatly affects device performance. The control width requirement will be more severe in the era of low-k material. Thus, easy film thickness control with end point is required.
3) Failure diagnosis/lifetime detection: CMP's reliability/reproductivity is low, and demands for developing end point control techniques have rapidly increased. It has been revealed that there are patterns in end point reliability or reproductivity. These patterns can be obtained by analyzing polishing stop and film thickness control information. For example, comparing end point data to such information can provide failure diagnosis and lifetime prediction.
4) Raw process time reduction: The figure shows how raw process time is reduced by using built-in polishing, cleaning, and end point detector (EPD) subsystems, instead of using standalone ones. The all-in-one type reduces the raw process time to about 1/3 compared to the conventional configuration using standalone subsystems. The portion of polishing time in the total process time increases, meaning that the use of send-ahead test wafers is easier despite the existing trend of omitting it.

Other control information

The figure shows a CMP control method including polishing, cleaning, transfer, slurry supply, and waste treatment. These parameters are critical to manage plants via higher-level communication. In-process monitoring is also necessary for processes other than polishing; scratches are related to the completeness of slurry supply, and the particle count is related to cleaning and transfer. Consumables also largely affect polishing characteristics. The polishing of new materials will involve environmental load reduction, including waste treatment.

Raw Process Time improvement

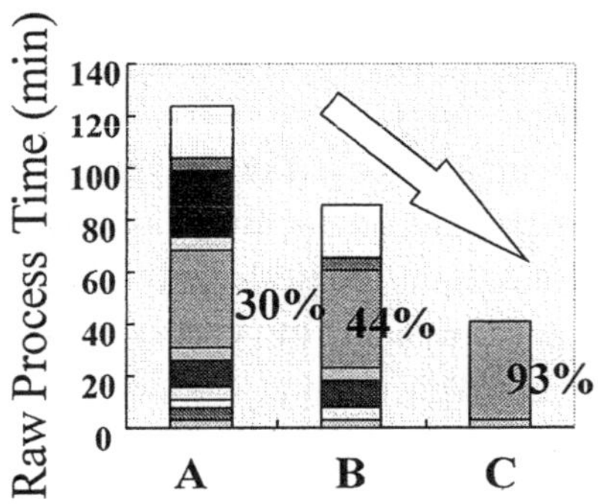

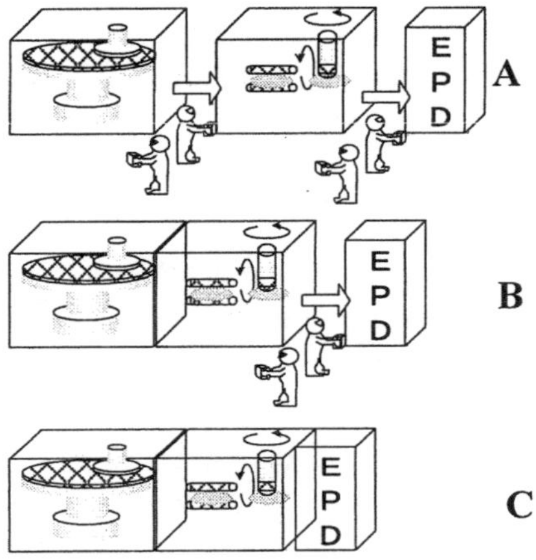

A : Stand alone
B : Dry-in dry-out
C : All in one

Communication to host

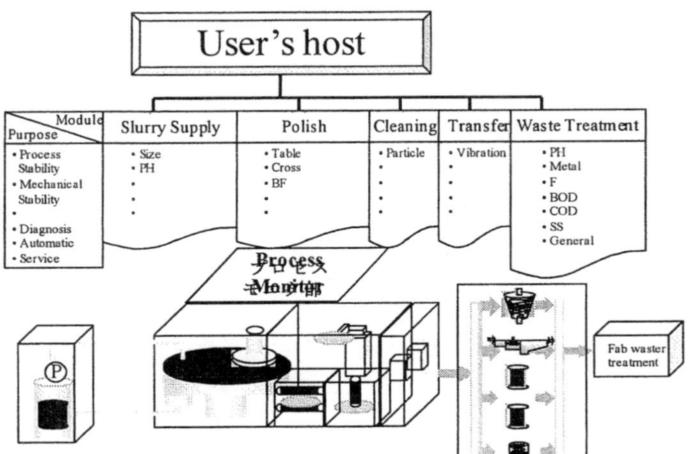

Slurry recycling and waste treatment

Slurry

Slurry performance depends on the *machining and chemical reaction of film to be polished with abrasives and polishing solution.* Oxide film is typically polished with KOH- or ammonia-based silica slurry, ceria slurry, or manganese slurry. The performance of such slurry is evaluated based on the polishing rate, scratches, and ease of removal. On the other hand, metal film is polished with hydrogen peroxide-, ferric nitrate-, or potassium iodate-based alumina/silica slurry or manganese slurry. The performance of such slurry is evaluated based on the polishing rate, scratches, ease of removal, dishing, oxide erosion, and corrosion.

An immediate challenge is to provide a new type of slurry for metal film polishing at low cost that meets the requirements above while maintaining selectivity appropriate for each film type (barrier film, underlying film, etc).

Slurry supply system

Slurry supply systems are divided into two types: on-site supply and central supply. The on-site supply type is suitable in the testing stage or if the slurry is frequently replaced. The central supply type is used in most actual production plants. Chemicals for other wet processes, such as plating, can also be handled with similar supply systems.

Slurry reuse and recycling

At present, the cost of slurry is extremely high, and slurry reuse/recycling has been considered. As shown in the figure on the next page, slurry is reused by circulating non-diluted used slurry through a filter. This method is generally used for the polishing of bare silicon wafers.

Recycling means that slurry is restored to its original condition. While there have been cases semiconductor processing chemicals are recycled on-site, slurry recycling is rather difficult. Currently, it is desired from an environmental perspective to achieve the recycling of slurry.

Chemical supply systems for plating and CMP

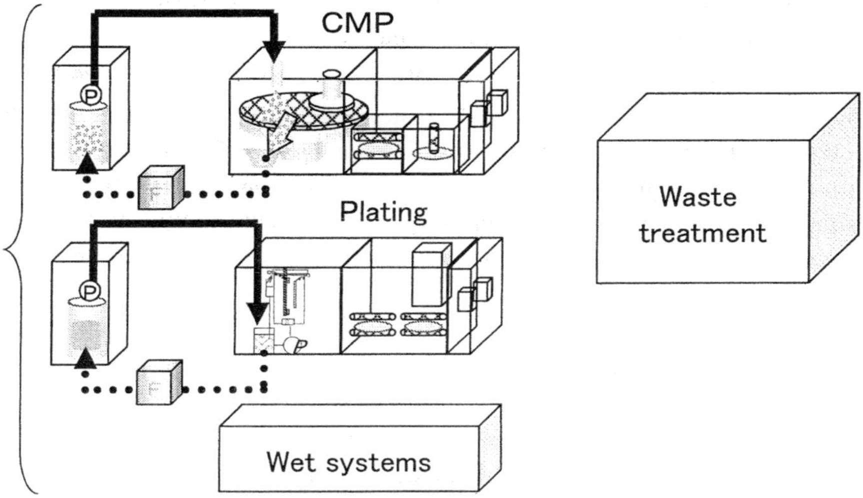

Slurry re-use / re-cycle system

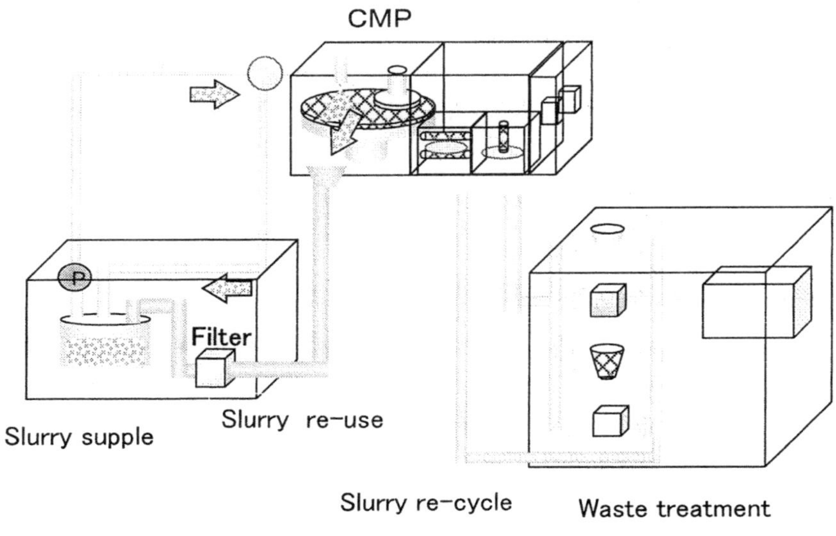

Let's start with ILD

Use of ILD CMP and definition of planarization performance

ILD CMP is the first adopted CMP process and is the most common type of CMP. When oxide film is CVD-deposited over Al interconnects as shown in the figure on the next page, a step height is generated. Despite planarization in the deposition process described before, step height generation is still inevitable. The step height generated upon deposition is defined as an initial step height and is minimized through polishing for planarization. The remaining height after polishing is defined as the final step height. The volume polished to reduce the step height to the final level represents planarization performance, as expressed below.

Planarization performance = (Initial step height – Final step height) ÷ (Polishing volume) (x 100 if expressed in %)

The 45-degree slope in the graph shows that *the step height reduces as the polishing volume increases*. It indicates theoretically 100 % planarization. The horizontal line denotes that *the step height does not reduce regardless of the polishing volume*. It indicates 0 % planarization. This concept of planarization performance may apply to wet etching, etc. The definition above is very fundamental and must be learned since it is important for discussing electro-chemical polishing (ECP) and other planarization technologies.

For dielectric film deposition, efforts have also been made to improve the gap filling capability. At present, film deposition with a substantially smaller initial step height can be achieved.

Increase in the planarization length!

The improvement of the gap filling capability provides a disadvantage. If the interconnect line and space (L&S) is wide, oxide film is somehow deposited along the interconnect pattern regardless of the gap filling capability. With a narrower interconnect L&S, however, dielectric film deposition along the interconnect pattern is impossible. The entire area over the interconnect pattern and its surrounding region is covered with the deposition.

This situation is highly problematic for CMP; the area to be polished is very wide, e.g. about 10 mm, which is equivalent to the chip width!

Such an increase in the area to be polished is a challenging problem for polishing. It is technically called an increase in the planarization length (PL). Let's proceed to the next section for a solution to this problem.

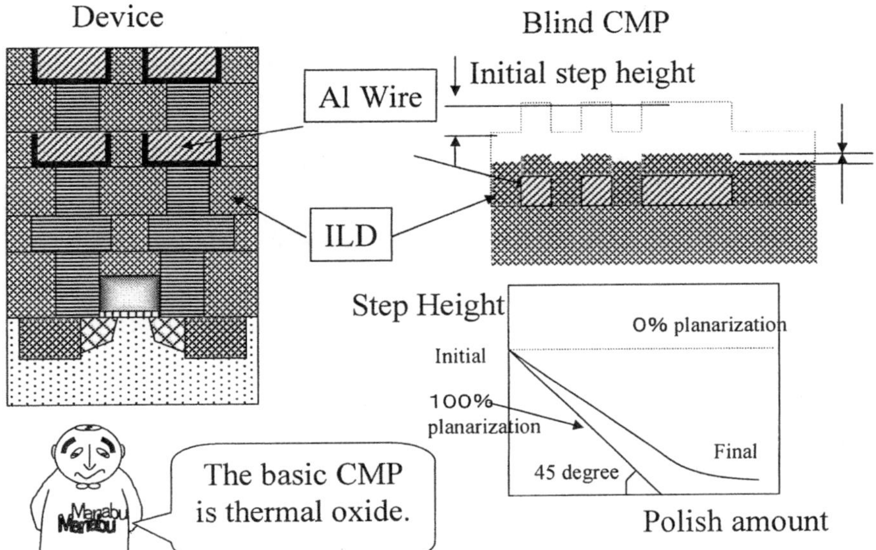

Planarization performance

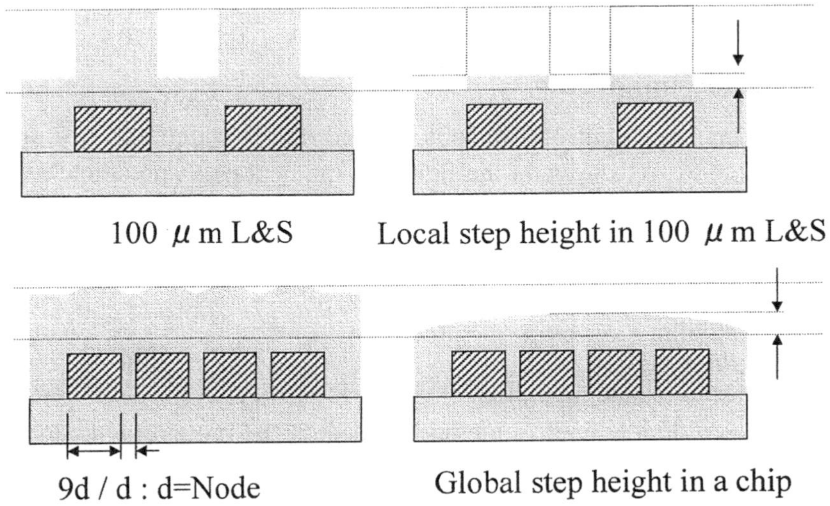

The more scaling, the longer planarization length

Profile control for uniformity upgrade

Principle of deplating

The principle of oxide film (SiO_2) polishing is as follows.

[Chemical] -OH groups on the surface are attacked by alkali to form a gel layer.

[Mechanical] The gel layer is mechanically removed.

Polishing rate and planarization performance

Oxide film polishing is the base of CMP, and Preston's equation is presented again.

Preston's equation: PR = kPV

The temperature effect (Arrhenius equation) is introduced here.

Modified Preston's equation: $PR = kPVT^{\alpha}$

As describe above, the dependency of slurry and pad data on PVT must be determined.

Planarization performance is calculated as follows.

Planarization performance = (Initial step height – Final step height) ÷ (Polishing amount) (x 100 if expressed in %).

In-wafer non-uniformity

Parameters and control methods related to in-wafer non-uniformity are explained below. As shown in the figure on the next page, parameters that affect wafer uniformity include:

1) Pressing load
2) Relative speed
3) Temperature
4) Slurry/pad type
5) Accuracy of table operation.

In the next section, control methods based on the pressing load are explained assuming that the parameters 2) to 5) are constant. There are the following types of control methods.

1) Control of the polishing rate profile with backside pressure
2) Control of the polishing rate profile with retainer ring load
3) Control of the polishing rate profile at the head edge/center

Details are given on the next page.

Principle of oxide CMP

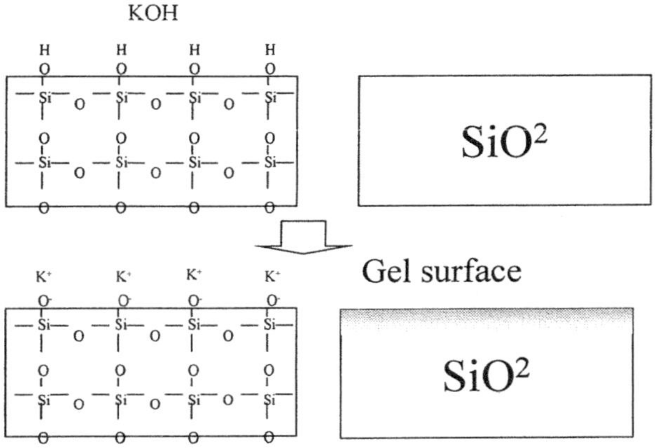

Control of wafer non-uniformity

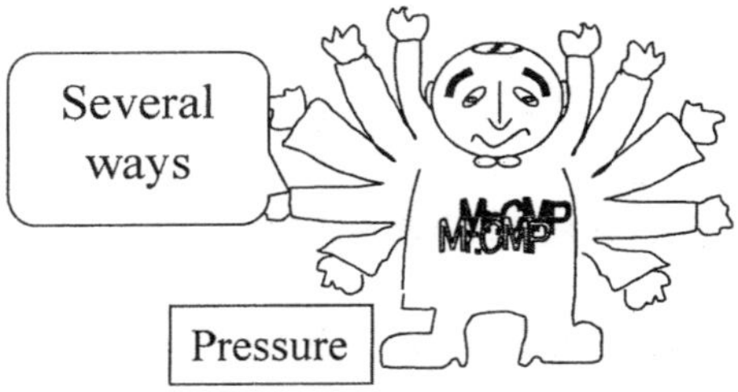

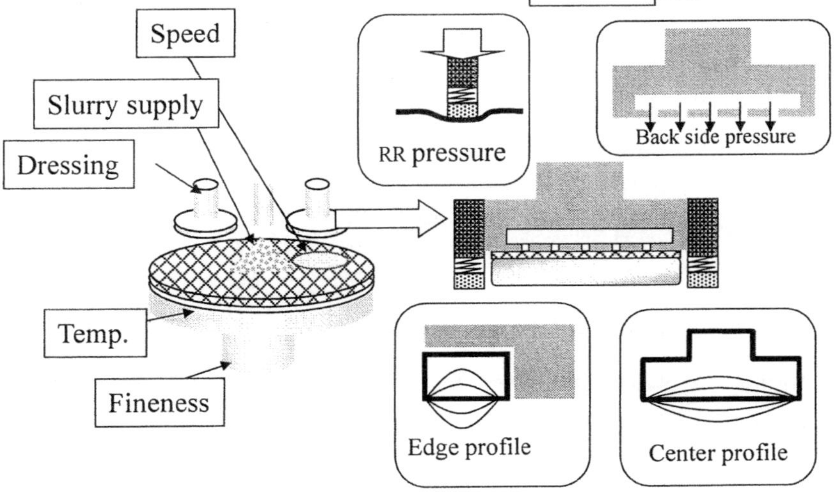

Control with backside pressure, edge pressure, and rotational speed

Three typical methods for controlling the polishing rate profile are explained.

Control of the polishing rate profile with backside pressure

The figure on the next page shows an example of polishing rate profile control with backside pressure. This method has been available since the emergence of CMP. It uses a porous wafer carrier to apply fluid pressure from the backside of the carrier for load control. Because of its simplicity and effectiveness, this method is still employed for laboratory machines.

This method is effective when the level of polishing uniformity required for CMP is low. However, it proves to be a bottleneck as a higher level of uniformity is required, and the capability of polishing systems is improved accordingly. Also, each wafer pad or slurry has its own peculiarities in polishing as described above; it is difficult to adjust pores on the pad to such peculiarities. Just as the saying goes that necessity is the mother of invention, various improvements have been made as outlined below.

Control of the polishing rate profile with retainer ring load

This method uses a ring designed to apply load to the wafer circumference. It can adjust pad rebound to control the polishing rate at the wafer edge. This method is used for the latest CMP systems and has been the first in the world to suggest that pad rebound must be considered. This control method is increasingly complicated with a variety of pad stiffness levels and thicknesses. The figure shows a schematic diagram of this method and the result of FEM analysis.

Control of the polishing rate profile based on the rotational speed

The figure also shows examples of polishing rate profile control based on the rotational speed. Typically, the same rotational speed is used for the table and the head so that the relative speed between the wafer and the pad is same. The central example shows the vector in such a case. The left and right examples show the vectors in cases the table speed is higher than the head speed and vice versa, respectively. This method allows you to control different polishing rates for the wafer edge and the wafer center.

Profile control by back side pressure

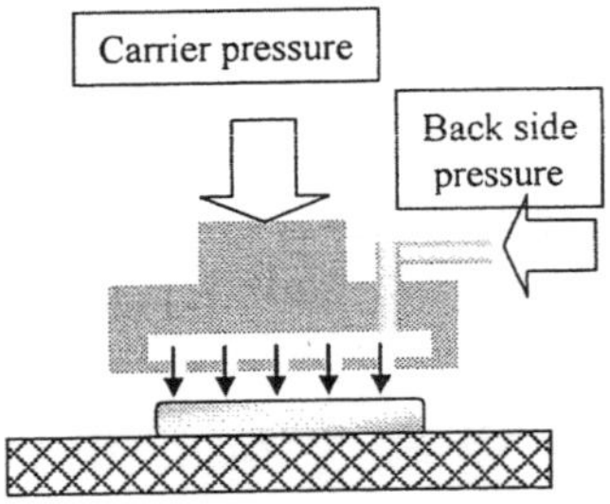

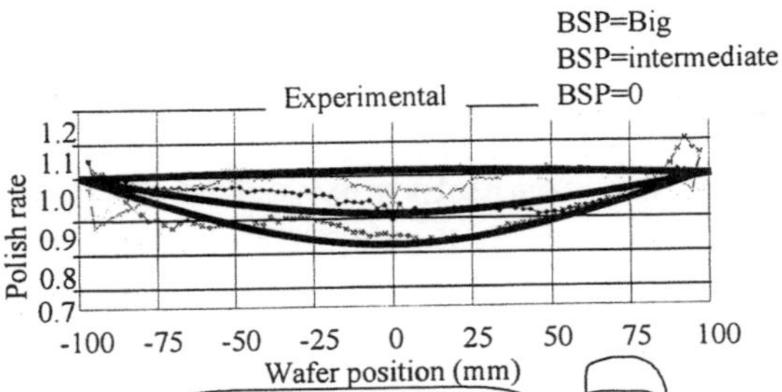

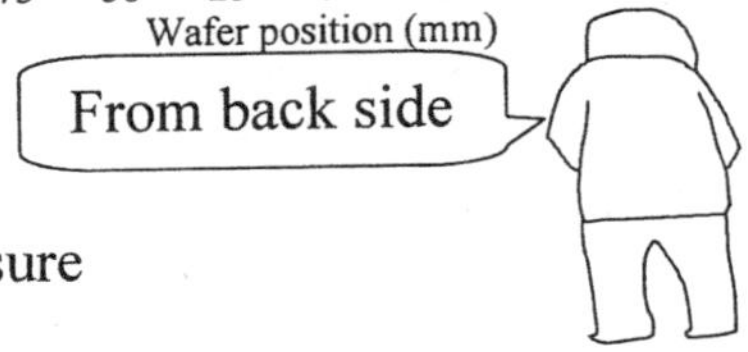

Profile control by retainer ring pressure

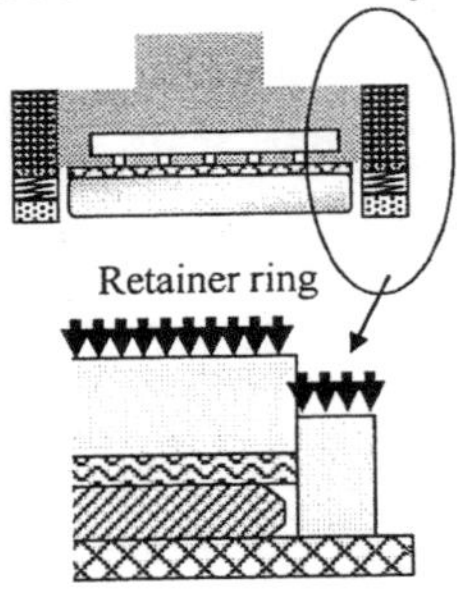

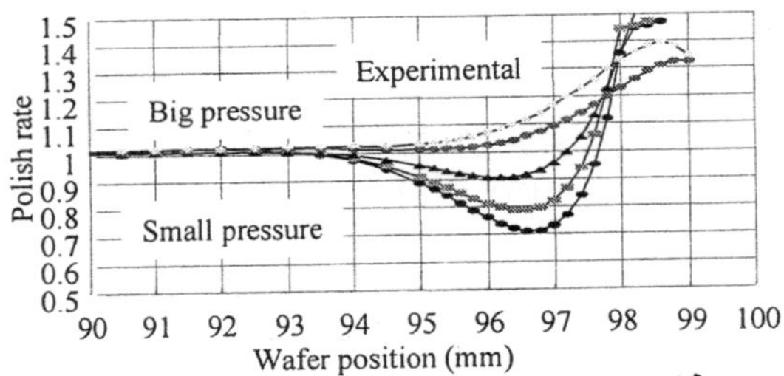

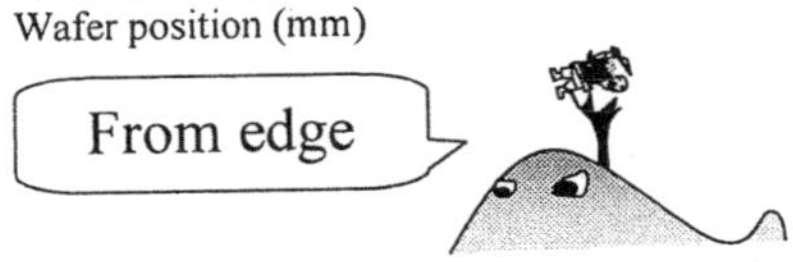

Profile control by rotation

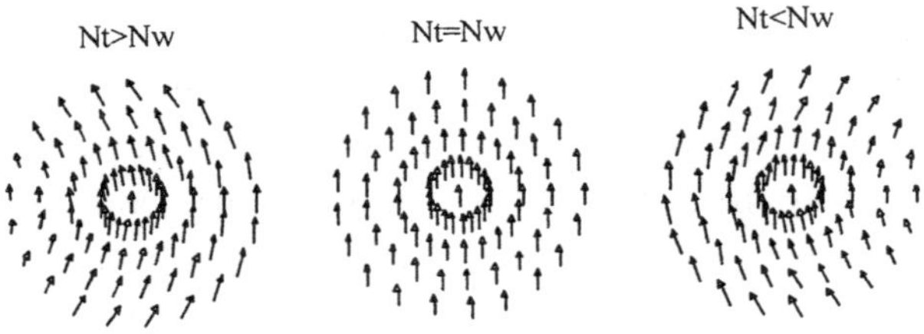

Note) N_T: Table rotation Nw: Wafer rotation

Dressing, flow, and vibration

Factors that affect in-wafer non-uniformity and appropriate measures for them are described below.

Use of a pad dresser

In-wafer non-uniformity deteriorates if the pad surface profile is not controlled with a dresser. The figure on the next page shows pad surface profiles made by different dressers. The pad surface profile can be controlled according to the dresser type and the dressing method. Depending on the process, the profile may be adjusted to be planar, slightly convex, or concave.

Slurry flow control

It is desired to minimize the slurry flow, but its limit of reduction must be identified. In the early period of the use of CMP, the slurry flow was adjusted within a range to obtain a constant polishing rate shown in the figure. For example, the standard slurry flow for 200 mm wafers was about 200 cc/m. Basically, the slurry flow required for 300 mm wafers is about 400 cc/m; because the cost of slurry is high, efforts have been made to reduce the slurry flow as much as possible. The figure shows the result of a test performed to determine the tolerance of the polishing rate to slurry flow reduction.

Stick-slip

Polishing conditions may significantly vary depending on the wafer, slurry, and pad types. For example, a polishing machine may vibrate when operated under polishing conditions that are apt to cause sticking (high friction). A possible cause of vibrations is the stick-slip phenomenon. The stick-slip phenomenon can be explained by supposing that wafers *get stuck* due to friction. To address this phenomenon, vibration analysis is performed to take anti-vibration measures, which will allow polishing in a broader range of conditions.

Improvement by pad dresser

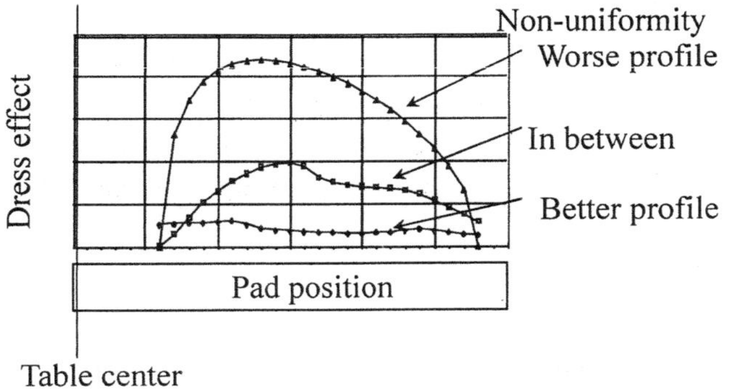

Slurry flow control

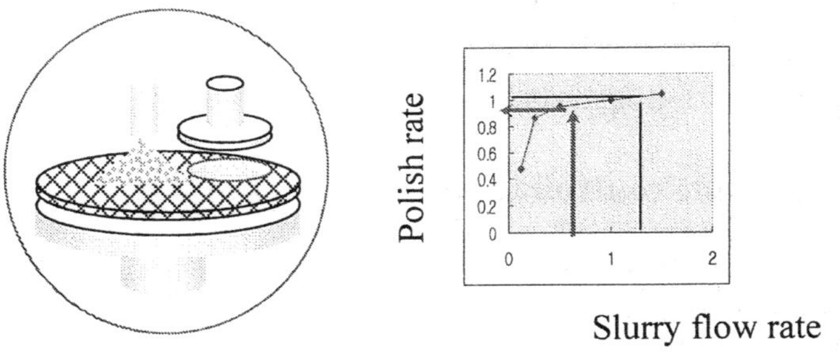

Stick slip

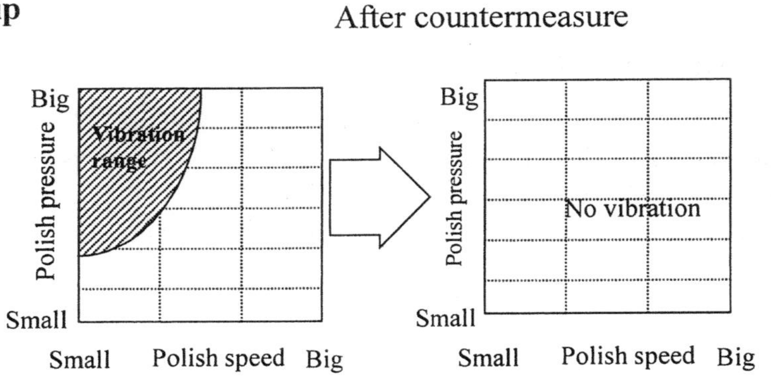

ILD specifications

The specifications of ILD CMP are introduced below.

Specifications

The figure on the next page shows the typical specifications of ILD CMP.

The polishing rate, in-wafer non-uniformity, planarization performance, and number of defects are specified as basic parameters. Other user-specific requirements, such as profile controllability, may be additionally specified.

In-wafer non-uniformity

The figure shows a typical profile of in-wafer non-uniformity.

Film: PE-TEOS

Pad: IC1000^{TM}-050 (K)/S400

Slurry: SS-25 (1:1)

Rotational speed of the table/carrier: 70/70 [rpm]

In actual operation, the stability of in-wafer uniformity must be demonstrated generally by performing a continuous test.

Polishing profile controllability

The correction of in-wafer non-uniformity profiles formed in the deposition process may be required before polishing. In such cases, the controllability of the polishing profile may be specified as a requirement. The profile control methods are as described above.

Step height reduction

The graph shows the curve of step height reduction with typical silica and ceria slurries. The degree of step height reduction largely depends on the device pattern. The pattern design determines locations that are difficult to planarize; thus, it is essential to well check the polishing result with the design pattern.

Specification

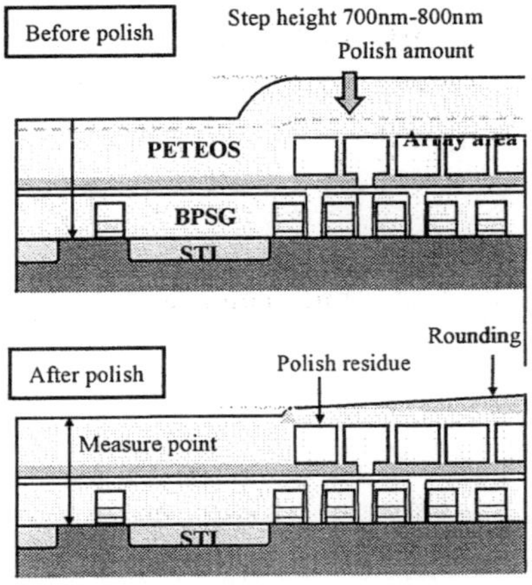

Performance (300mm wafer)

Items	Requirements
Polish rate	≧300nm/min
Non-uniformity(WIW)	≦3% (1σ) EE:3mm
Defect (Surfscan6420)	≦30 pcs(≧0.2μ)

Example of wafer non-uniformity

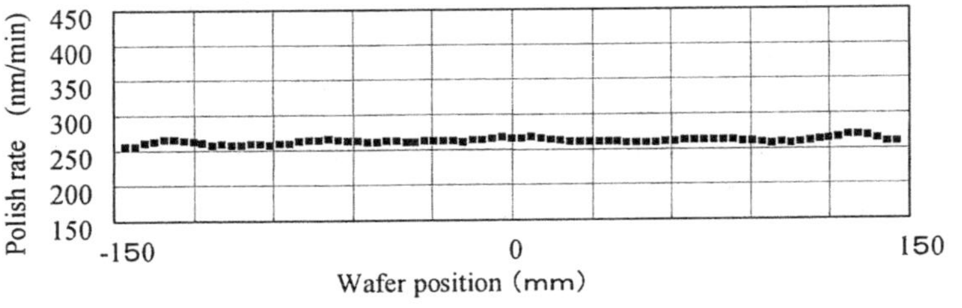

Example of step height reduction

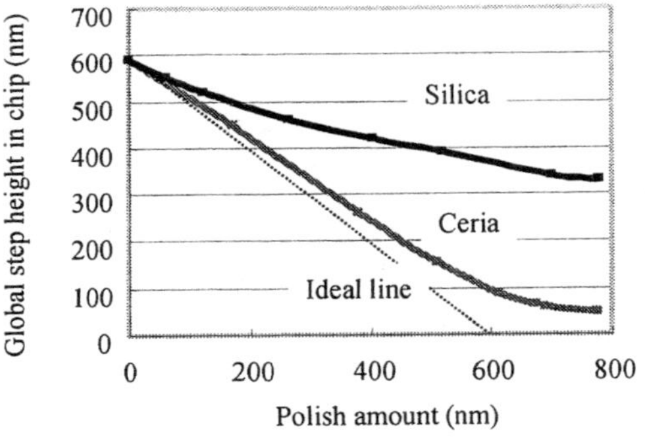

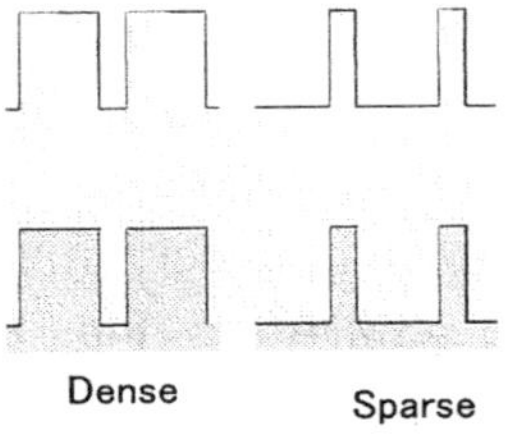

STI CMP

Use and problems

The use and performance of STI CMP are shown in the figure on the next page. STI CMP has problems related to the selectivity of stopper film (Si-N) and defect reduction. The photos show STI before and after polishing.

Principle

STI CMP mainly uses silica slurry or ceria slurry. The principle of polishing with silica slurry is the same as that of ILD CMP.

In-chip non-uniformity

As a special requirement for STI CMP, the selectivity of stopper film (Si-N) comes to an issue. Typically, the selectivity of SiO_2 film with respect to Si-N film is made high to reduce the polishing of Si-N film. A higher selectivity of SiO_2 film with respect to Si-N film means that the SiO_2 polishing rate is higher than the Si-N polishing rate, generally resulting in a larger dishing amount of SiO_2 film (trade-off relationship). Overpolishing is a key to solving this problem; STI CMP should be performed by (1) increasing the selectivity of Si-N film with respect to SiO_2 film (1:30, for example) and (2) ensuring that the dishing amount of SiO_2 is suppressed even with overpolishing.

Defects

There is no doubt that the most critical problem for STI CMP is scratching. Any scratch passing through Si-N film is fatal and not permitted. In addition, any scratch on SiO_2 film that does not pass through it is undesirable. The ratio between the thickness and depth of scratches developed before Si-N film deposition is important. Suppose a scratch of 10 μ in width and 1 μ in depth. It is important to determine if such a scratch is transferred to the final stage. If so, the scratch develops into a critical problem; you will face a challenge of planarizing the scratch with a width/depth ratio of 10:1.

Monitoring

Highly selective polishing with ceria slurry allows end point detection based on frictional force.

Device image

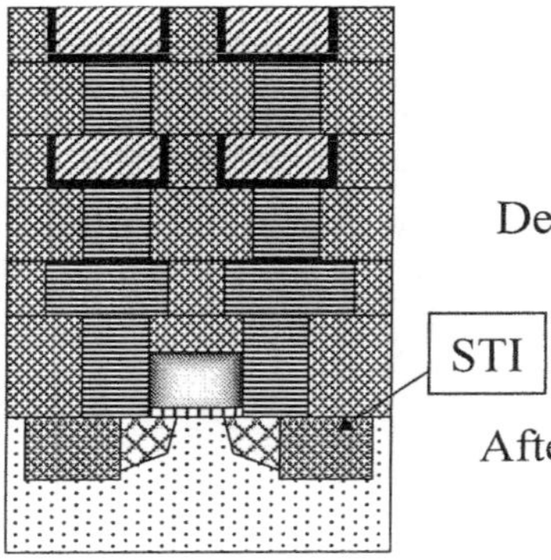

Recess CMP

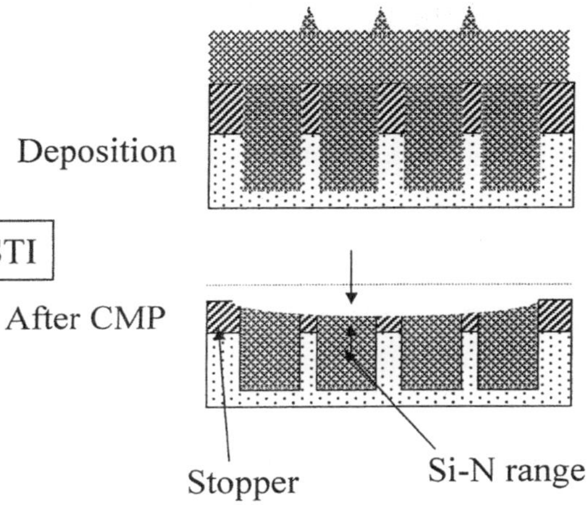

Example of STI

	Cell Array (Center)	**Isolated**
Before CMP	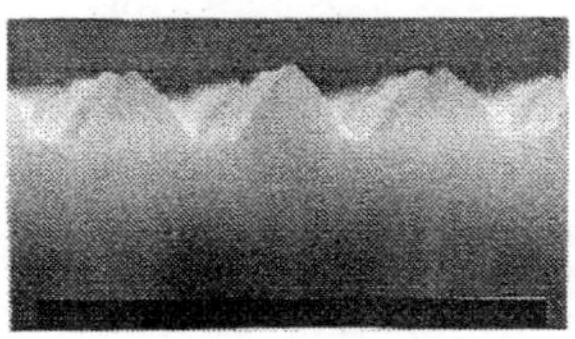	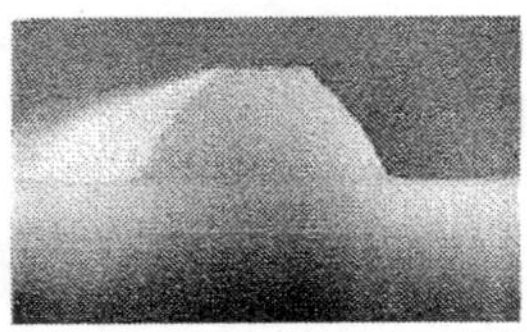
After CMP	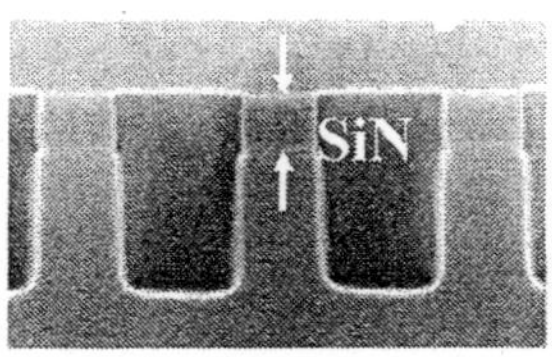	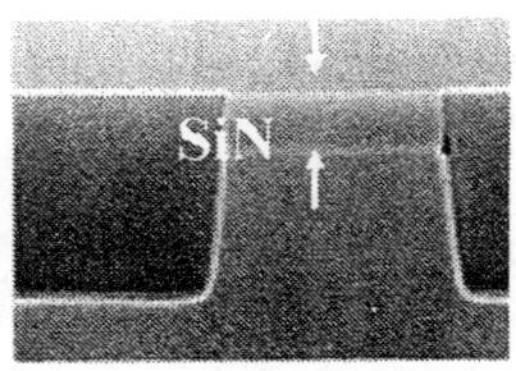

STI CMP methods

Various STI CMP methods are explained below with descriptions on how to maximize the use of STI CMP.

Reverse etching and direct polishing

Typically, direct CMP with no additional measure may result in an inability to planarize a wide field, consequently generating defects in stopper film. One way to address this problem is reverse etching. As shown in the figure on the next page, this technique pre-etches the field that is difficult to planarize. In this way, it facilitates planarization by decreasing the planarization length.

Dummy patterning

This technique forms dummy fills for patterns susceptible to dishing before polishing. The dummy fill layout and design are not detailed here since they are related to know-how and patents of device manufacturers. It should be added that the latest devices are full of dummy fills.

Ceria abrasive

There is a question about whether a simple CMP process without the measures above must always polish a large planarization length. To decrease the planarization length, various efforts were made, such as providing higher pad stiffness or increasing the polishing rate for increasing the effective stiffness of the pad. However, the effect of these efforts was insufficient; therefore, planarization with ceria slurry was proposed. In the present day, ceria slurry is commonly used for planarization.

Fixed abrasive

One way to provide higher pad stiffness is the use of fixed abrasives. Polishing with fixed abrasives is performed by fixing ceria particles in a pyramid form on a normal pad (as in the case of 3M Corporation) or by producing a pad made of ceria particles. The photo shows a pad made of ceria particles.

For the use of fixed abrasives, defect reduction must be addressed.

Reverse etching and direct CMP

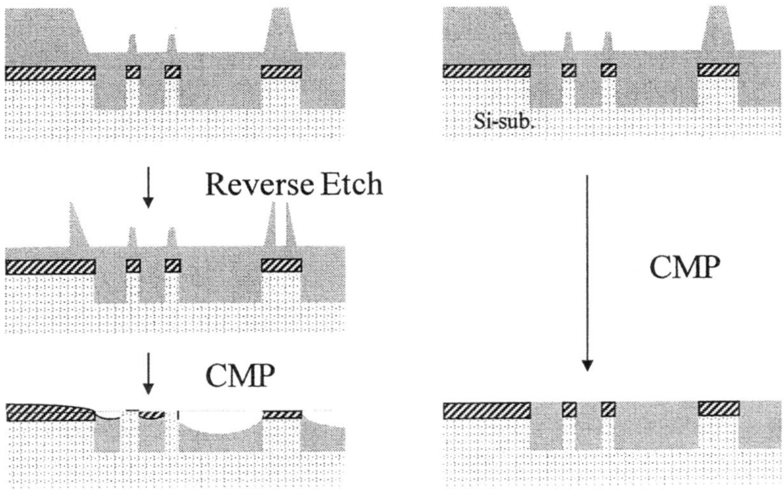

Reverse etch

Direct polish

Example of bonded abrasive

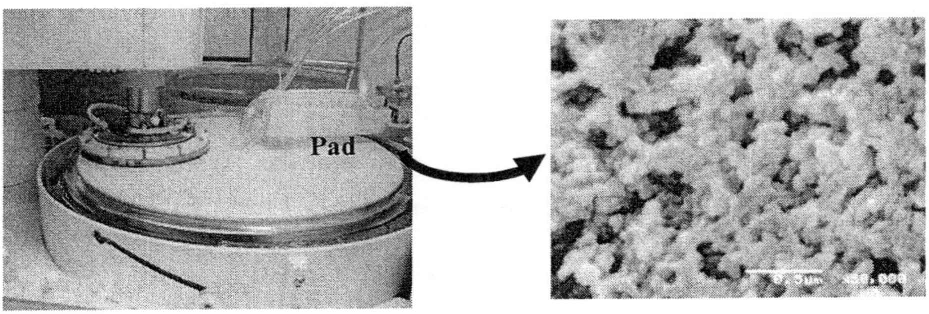

Bonded abrasive table

SEM

Tungsten (W) CMP

Use and problems

The figure on the next page shows a schematic diagram of a tungsten (W) contact and via. Typically, a dielectric layer is made of oxide film, and a barrier layer is Ti/Ti-N. Thus, W CMP involves the polishing and recess of different materials (W -> Ti/Ti-N).

For W CMP, the following parameters are specified (values depend on the generation).

Oxide recess : < 10 nm

Oxide erosion : < 20 nm

Dishing : < 10 nm

There is no doubt that smaller values of these parameters are preferable. To achieve this goal, two types of CMP processes are available: a two-stage CMP process, which consists of two separate steps of tungsten/barrier (beneath the oxide film) polishing and slurry use, and a single-stage CMP process, which polishes all materials at the same time. The single-stage process is superior in productivity while the two-stage process is superior in process performance. The single-stage and two-stage processes may be used separately according to the requirement of each generation.

Principle

The principle of W CMP is as follows.

(1) The W surface is oxidized with an oxidizer, such as hydrogen peroxide or ferric nitrate.

W + oxidizer -> WO_X

(2) The oxidized W film is mechanically removed. At this time, only the convex part is selectively removed with a stiff pad for planarization.

Device image

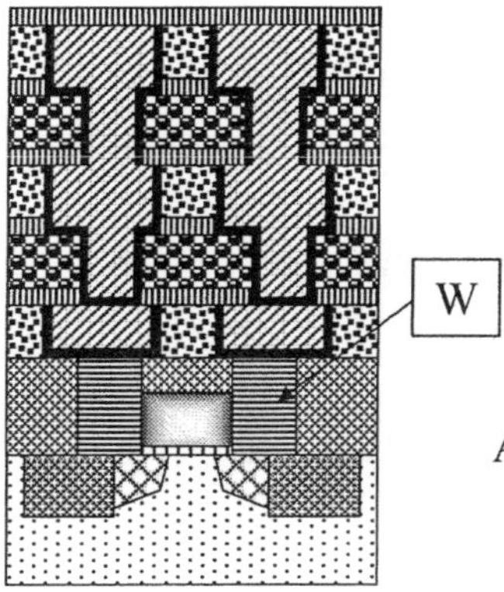

Recess CMP

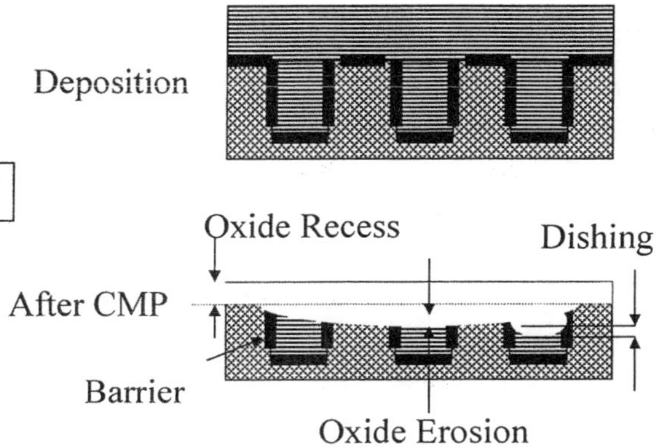

The principle of W-CMP

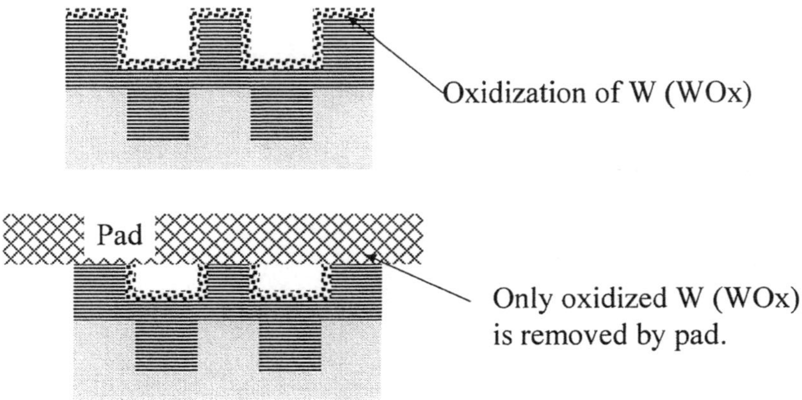

Cu residue due to W CMP of underlying layers

Two recent issues related to W CMP are introduced below.

Short circuit of Cu interconnects caused by erosion

CMP planarization of tungsten contacts may cause a short circuit of Cu interconnects on top of such contacts. Accordingly, planarization requirements for W CMP are recently becoming more stringent. The figure on the next page shows the mechanism of Cu residue formation.

(1) Tungsten deposition is performed for filling contact holes. It is conducted with CVD, providing a relatively conformal deposition.
(2) W CMP causes oxide erosion (schematic diagram).
(3) An oxide film is deposited conformally on the erosion shape.
(4) With the same height of etching, the etching shape is conformal to the erosion shape.
(5) Plating deposition is performed on the oxide layer. With mounding in the plating film ignored, the plating film is also conformal to the erosion shape.
(6) The shape of Cu polishing is conformal to the erosion shape.
(7) Consequently, Cu remains over the eroded region (when seen from above).

As described above, W CMP produces Cu or other metal residues on top of an oxide layer. To solve this problem, you must minimize erosion caused by W CMP and remove any residue by overpolishing.

Edge over erosion (EoE)

A higher selectivity of a specific material among different ones may result in a significant spiking phenomenon around the edge (edge over erosion: EoE). This problem is solved by adjusting the selectivity.

Mechanism of Cu residue

①W deposition

②W－CMP Erosion

③Oxide deposition

④Oxide etching

⑤Cu depositon

⑥Cu CMP

⑦Cu residue (upper view)

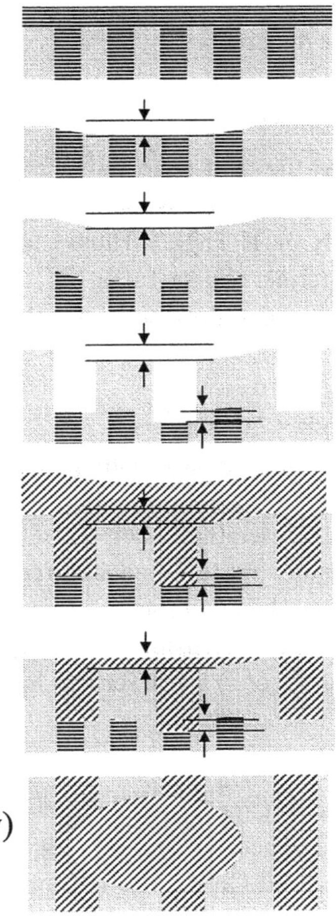

Edge over erosion (EOE)

Cu CMP

Cu CMP is explained below in the same way as for W CMP to compare them and clarify their differences.

Use and problems

The figure on the next page shows a schematic diagram of Cu interconnects. Typically, a dielectric layer is made of oxide film or low-k material, and a barrier layer is Ta/Ta-N or Ti/Ti-N. Thus, Cu CMP involves the polishing and recess of different materials (Cu -> Ta/Ta-N).

For Cu CMP, the following parameters are specified (values depend on the generation).

Oxide recess: < 10 nm

Oxide erosion: < 20 nm

Dishing: < 30 nm Note) Dishing is larger since a larger pad is used.

As is the case with W CMP, there is no doubt that smaller values of these parameters are preferable. For Cu CMP, two types of processes are mainly used: a two-stage process and a three-stage process. With the adoption of low-k material, planarization requirements for Cu CMP (including damage specifications) are increasingly stringent. Performance requirements (lower dielectric constant, etc.) are described later.

Principle

The principle of Cu CMP is as follows.

(1) The Cu surface is oxidized with an oxidizer (valence increase).

(2) The oxidized Cu surface is (immediately) complexed with a complexing agent.

(3) The Cu complex is mechanically removed. At this time, only the convex part is selectively removed with a stiff pad.

$Cu \dashrightarrow Cu^{2+} + 2\,e^-$

$O_2 + 2\,H_2O + 4\,e^- \dashrightarrow 4\,OH^-$

Cu + complex --> Cu (complex)

Cu (complex) ↑

For details on Cu CMP, see "General principles of planarization technologies" later.

Device image

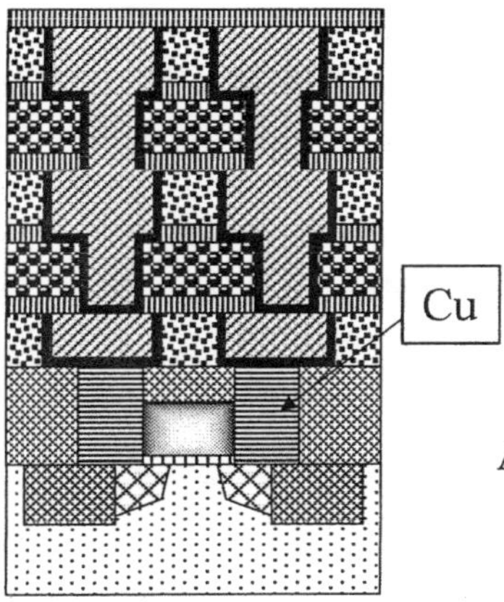

Recess CMP

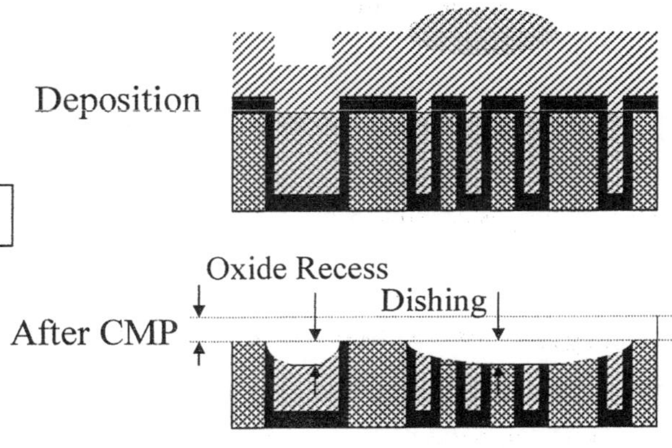

The principle of Cu CMP

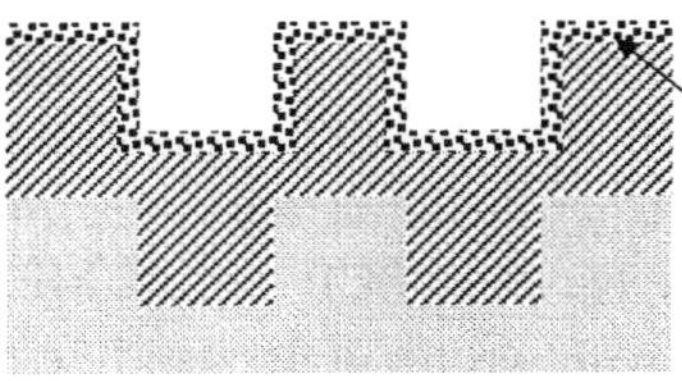

Cu is oxidized.
Cu + Oxidizer-> CuOx
Oxidized Cu become complex..
CuOx + (Complex) -> Cu (Complex)

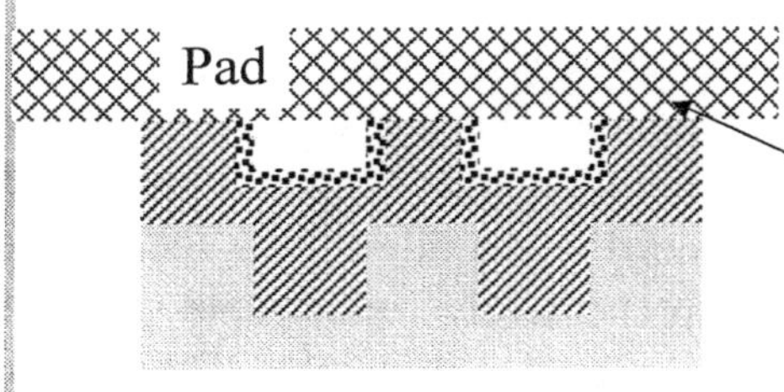

Only Cu complex is removed by pad.

General principles of planarization technologies

This section introduces cutting-edge technologies for damascene interconnect planarization for the 45 nm generation or beyond. It explains a special theory of general principles of planarization covering polishing (deplating) only and a general theory covering both polishing and plating (deposition). These principles produced a so-called mC^2 process, honorably named after Dr. Einstein's equation of the relativity theory. I hope that you have a sense of humor to enjoy it. The general principles, which have been popular with my students, are introduced below.

Emergence of various planarization technologies

In the late 1980s, polishing appeared as innovative planarization technology. It was first employed for ILD (SiO_2 film) planarization; then, it has been applied to STI, tungsten via, Al damascene, Cu damascene, etc. Recently, the possibility of its use for metal gates has been discussed. The approach to polishing has also evolved. Especially since, as ILD material, SiO_2 film was replaced by weak low-k material for a lower dielectric constant, various technologies have been introduced to reduce polishing pressure. Besides conventional CMP, the Electro-Chemical-Mechanical Polisher (ECMP), Electro-Chemical Polisher (ECP), Chemical Etching (CE), and their combinations are available now.

There is a question about which is the most promising among rapidly burgeoning technologies. Also, terminology for electrolytic polishing has not been integrated; some ask, "Are abrasives not used for electrolytic polishing?" Indeed, confusion exists between electrolytic polisher (without abrasives) and combined electrolytic polisher (with abrasives) or between ECP and ECMP in some articles and papers. It is natural that authors with wrong understanding are confusing their readers.

The background of demands for and the use of low-pressure polishing are explained with various types of low-pressure polishing presented. Then, a general explanation is given about all planarization technologies based on a general theory. The *general principles* mentioned in the title of this section are not actual, but aim to help you comprehensively understand various planarization technologies by clarifying their underlying principles. The general principles are greatly welcomed by my students. I am happy if the descriptions below provide good guidance for confused readers. The technologies of cleaning/drying after Cu/low-k material planarization are also presented.

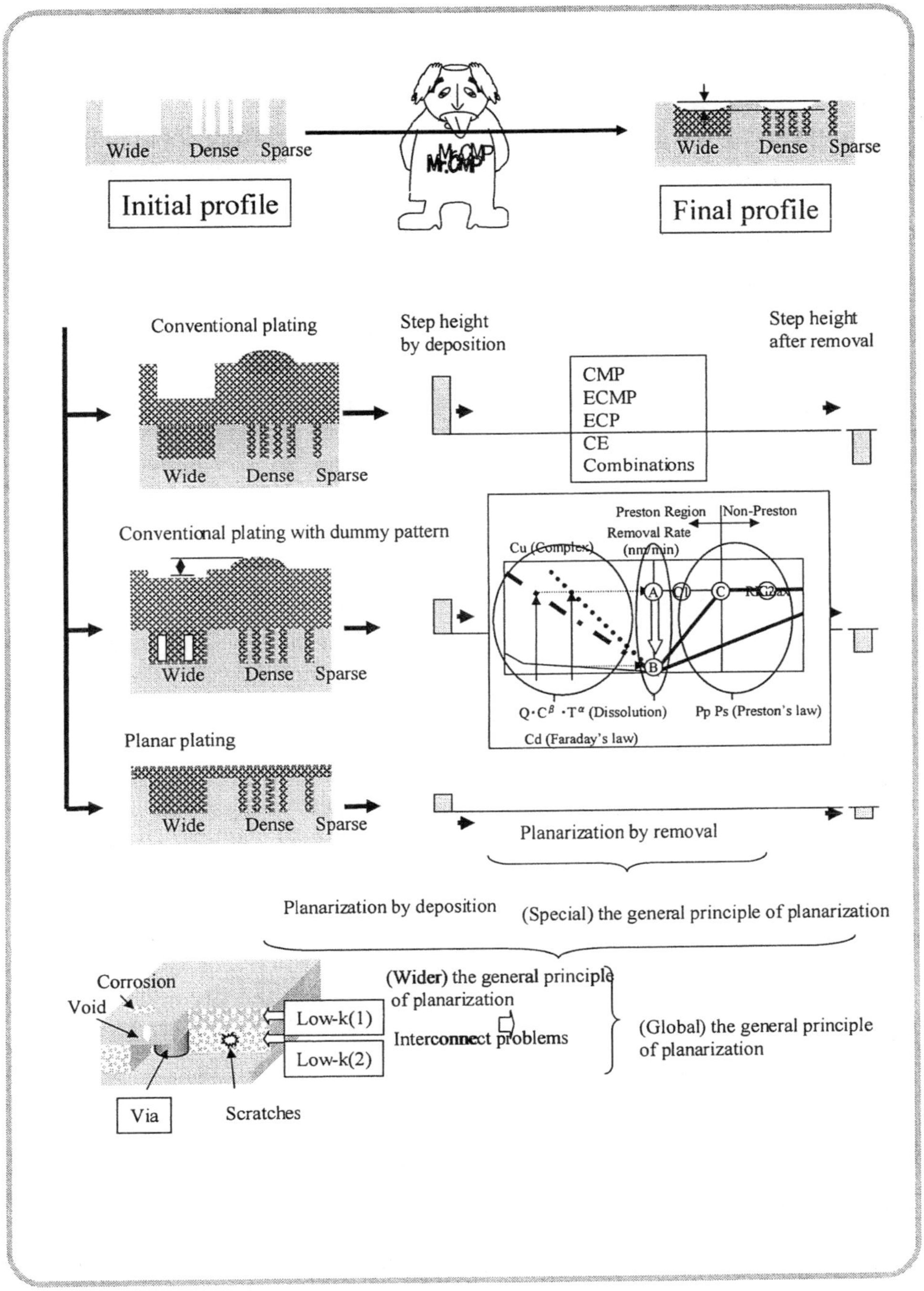

Wide
Dense
Sparse
Initial profile
Mr.CMP
Wide
Dense
Sparse
Final profile
Conventional plating
Step height
by deposition
Step height
after removal
CMP
ECMP
ECP
CE
Combinations
Wide
Dense
Sparse
Conventional plating with dummy pattern
Preston Region
Non-Preston
Removal Rate
(nm/min)
Wide
Dense
Sparse
Q·C^β ·T^α (Dissolution)
Pp Ps (Preston's law)
Cd (Faraday's law)
Planar plating
Wide
Dense
Sparse
Planarization by removal
Planarization by deposition
(Special) the general principle of planarization
Corrosion
Void
Low-k(1)
Low-k(2)
Via
Scratches
(Wider) the general principle of planarization
Interconnect problems
(Global) the general principle of planarization

Lower and lower, so lower?

Why is low-downforce polishing required?

As well known, the interconnect material changed from Al to Cu to reduce the resistance-capacitance (RC) delay of semiconductor devices. To reduce the dielectric film capacitance, many low-k materials have emerged as candidates. A lower dielectric constant proportionately decreases film strength and hardness. Adhesion to underlying film must also be considered. To solve these issues, low-downforce polishing has been proposed. The figure on the next page shows the evolution of device processing and a damascene structure using Cu and low-k materials. As low-k material, plasma-treated SiO_2 film (substrate), F-doped SiOF film, SiCOH film (replaced by $-CH_3$), and such films with pores are commonly used. The dielectrics of interconnect and via in the figure are made of different low-k materials, but the same material may be used. No etch stopper (ES) may be required. The dielectric cap layer (DCL) is made of Si-N, Si-C, or Si-CN.

How low is the downforce for polishing?

How low is the required downforce? Let's look at the physical characteristics of low-k material. The graph shows a relationship between the dielectric constant and Young's modulus. The strength of SiO_2 and SiOF films is about 70 and 40 GPa, respectively. Most low-k materials, however, show the strength of 10 GPa or less. Of course, processes other than polishing also cause film stress; it is known that thermal cycles and bonding for implementation give higher film stress. Film stress in one process does not cause fatal damage (if so, such a situation is out of the question), but it is like fatigue gradually accumulated through processes. The total stress caused by all processes should be reduced, and lower downforce for polishing is required in this context.

Strong low-k materials have also been developed (10 GPa or more). For those with a strength of about 5 GPa, a process for increasing the strength with UV or electron beam curing has been developed.

Young's modulus is a measure of breakdown and fracture strength of Cu/low-k damascene structures. Hardness is a measure of resistance to scratches that pose a critical problem to polishing. Generally, materials with lower Young's modulus have lower hardness. Adhesion is also an important factor since delamination is a critical problem in polishing. It is typically measured with a 4-point bending technique; it is often about 5 to 10 J/m^2, but 10 J/m^2 or more is desired. A process for increasing the adhesion has reportedly been developed.

Scaling and damascene integration

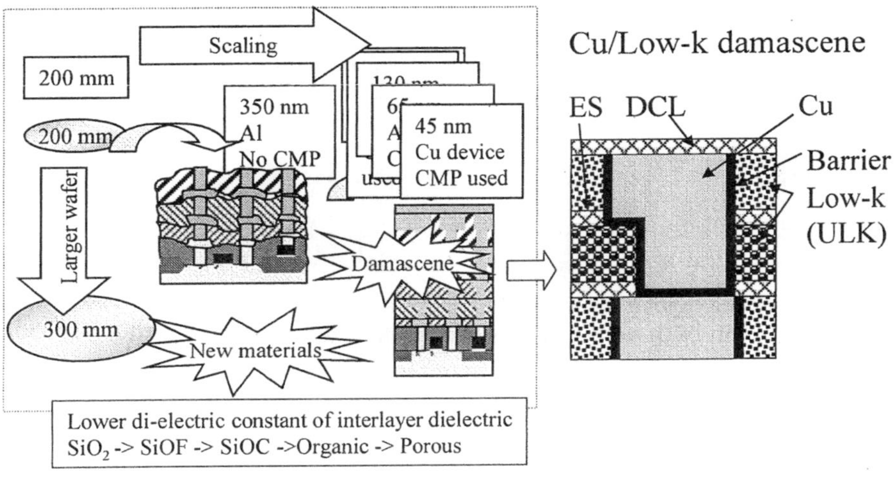

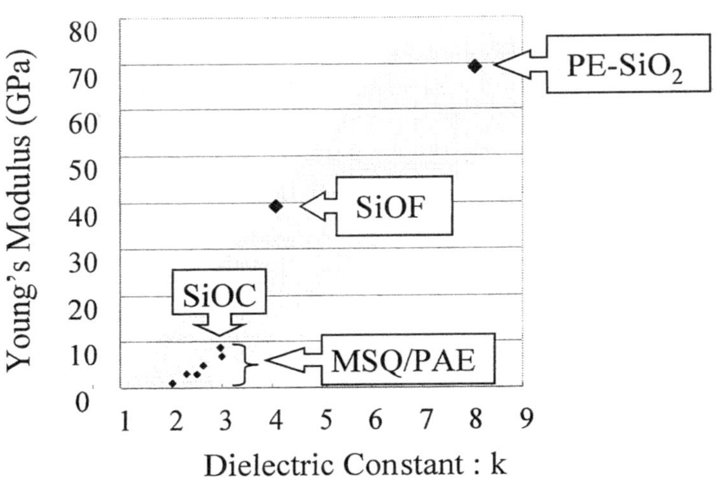

Di-electric constant vs young modulus

So, let's prove with FEM.

Here again, FEM can be used to determine a relationship between the pressure and the damascene structure condition. For the analysis of low-k material, fracture mechanics must be applied in addition to stress analysis presented below. Stress field analysis can be directly used, but there is a difference whether stress or fracture energy is used as a judging criterion.

Stress analysis

FEM is effective for analyzing material strength and adhesion.

The figure on the next page shows the analysis diagrams of the damascene regions. This analysis is performed separately for the interconnect region and the low-k region. For both regions, the analysis result shows that the via root is subject to the highest stress.

Analysis results

The graph shows the result of stress analysis using a damascene structure predicted in the ITRS. The x-axis indicates the effective dielectric constant calculated based on the estimated dielectric constant of each low-k material. The y-axis indicates the stress ratio, σ/σ (90), for Cu interconnects and the inside of the low-k materials. σ (90) represents stress to be generated when a damascene structure of the 90 nm generation is polished at 0.5 psi. σ/σ (90) represents a ratio of stress for the damascene structure of each generation to σ (90). For reference, the value of stress upon polishing a 90 nm generation damascene at 2 psi is indicated as an acceptable value. For both the Cu interconnect region and the low-k region, the via root is subject to the highest stress, the value of which is taken as σ. This data may be helpful for estimating polishing pressure for the damascene structure of each generation.

Low-k material has lower strength and may cause delamination, deformation, and scratches due to polishing. As a solution to these concerns, low-downforce polishing has been demanded. However, low-downforce polishing necessarily involves a lower polishing rate, increasing the CoC. A balancing point (happy middle ground) between these factors must be found.

The result of fracture mechanics analysis is given for reference. In terms of fracture mechanics, the strength of low-k material is important, but the matching between vertically layered different materials is more critical. The result of fracture mechanics analysis agrees with that of FEM analysis except in case the matching is extremely improper.

FEM analysis of damascene structure

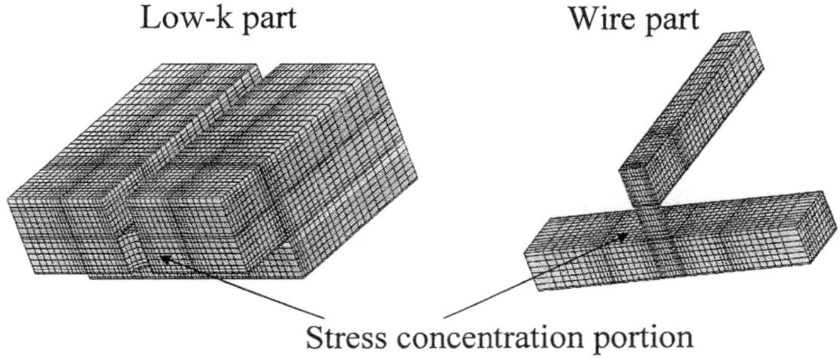

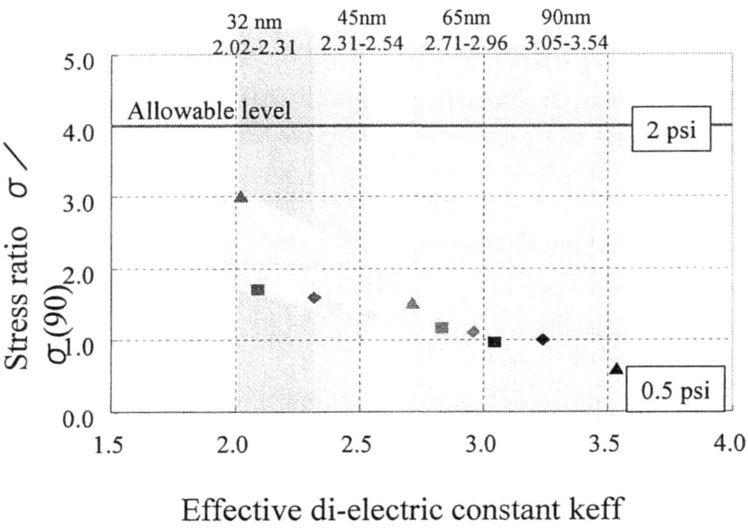

Di-electric constant vs CMP stress

Various planarization methods

Conventional planarization technologies

Conventional planarization methods are introduced below.

(1) Grinding

The method is used to grind the back surface of wafers, not the device surface. It may damage the surface, resulting in wafer defects.

(2) Fixed abrasives or bonded abrasives

The method is positioned between grinding and polishing. Fixed abrasives become free to grind wafers; the principle is the same as that of polishing. This method shows good planarization performance and is applied to some STI polishing. However, defect minimization is required.

(3) Chemical mechanical plolisher (CMP)

Cu CMP consists of 1) Cu oxidation by an oxidizer, 2) Cu complexation by a complexing agent, and 3) mechanical removal of Cu complexes.

The planarization technologies of the next generation have been born to respond to the demand for low-downforce polishing.

Applied electrolysis/etching and combined technologies

(4) Electro-chemical-mechanical polisher (ECMP)

Cu ECMP consists of 1) Cu oxidation by electric energy and some use of an oxidizer, 2) Cu complexation by a complexing agent, and 3) mechanical removal of Cu complexes.

The steps except 1) are the same as for Cu CMP; planarization performance is also similar to that of Cu CMP. ECMP faces challenges, such as developing a polishing pad that satisfies both electrolytic and planarization designs.

(5) Electro-chemical polisher (ECP)

ECP is a reverse step of plating and may be called deplating. It starts polishing (etching) when electrolytic concentration occurs for planarization. It does not offer the same level of planarization performance as polishing since it limits electrolytic concentration patterns. However, ECP is highly regarded with potential to inspire various innovations. For example, specialized ECP is developed to only etch locations that ion exchanges contact in ultrapure water.

(5) Chemical etching (CE)

Isotropic/anisotropic etching is a standard method for dry processes. CE performs anisotropic etching in wet etching processes. It is used in the MEMS wet process for some materials, but it has problems in device application.

Comparison of several planarization technologies

Process	Planarization length (mm)	Pressure (PSI)	Roughness
Grinding	>20	Big	Big
Fixed abrasive	~20	intermediate	intermediate
Polish	~2	0. 25~10	Small
ECMP	~2	0. 25~10	Small to intermediate
ECP	~2	0	Small to intermediate
Etching	0(~?)	0	Small to intermediate

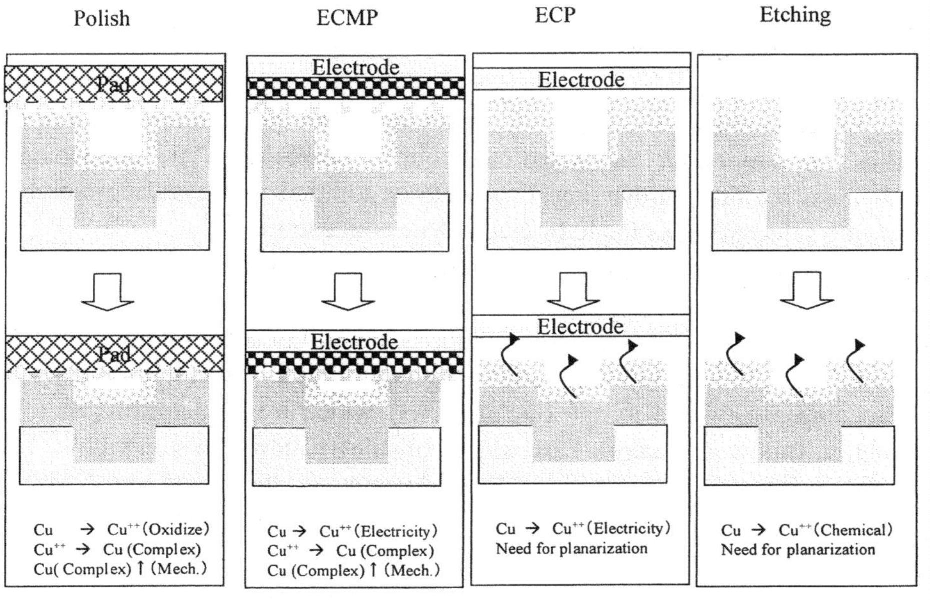

CMP/ECMP/ECP/CE: What do they mean?

Special theory of general principles of planarization

Promising low-downforce Cu polishing methods include CMP, ECMP, ECP, CE, and their combinations. The figure on the next page shows their principles. Based on the principles of such technologies, their features are explained.

CMP principle (Preston's Law)

First, how to interpret the graphs is explained. The left part of the x-axis shows a driving force for removing and planarization mechanism; the right part shows the product of a downforce Pp (kPa), often seen in CMP, and a relative speed Ps (m/s). The y-axis shows a polishing rate (nm/min). Point A indicates a driving force for removing; B indicates a state with the force controlled. In CMP, Cu is oxidized to Cu^{++} by an oxidizer (Point A), and Cu^{++} is converted to Cu complexes with a complexing agent (B). Cu complexes are then mechanically removed, the rate of which is proportional to the polishing pressure (Pp) and relative speed (Ps). This complies with Preston's Law, but the polishing rate relies on the Cu complex type.

ECP principle (Faraday's Law)

The left part of the x-axis shows a current density Cd (A/cm^2), and the driving force for removing is assumed to obey Faraday's Law. This is the reverse of film plating. A suppressor is added to control removing. The planarization mechanism requires further improvements in addition to the core principle of removing selected peaks made by electrolytic concentration.

ECMP principle (Faraday's Law and Preston's Law)

The left part of the x-axis shows a current density Cd (A/cm^2), and the driving force for deplating is assumed to obey Faraday's Law. Cu is oxidized to Cu^{++} by electric energy (Point A). Following the same process as CMP, Cu^{++} is converted to Cu complexes with a complexing agent (B), which are mechanically removed. Oxidation follows Faraday's Law while polishing follows Preston's Law. Thus, the entire process, except oxidation method, is the same as CMP; its planarization mechanism can be said to be the same as for CMP.

CE (dissolution law)

The left part of the x-axis shows the product $Q \cdot C^{\beta} \cdot T^{\alpha}$ of the etchant's flow (Q), concentration (C), and temperature (T). The driving force for removing is assumed to obey the dissolution law. A suppressor is added to control removing. The entire wafer surface is dissolved with the etchant; CE has no planarization capability. Additional chemical agents selectively removing peaks.

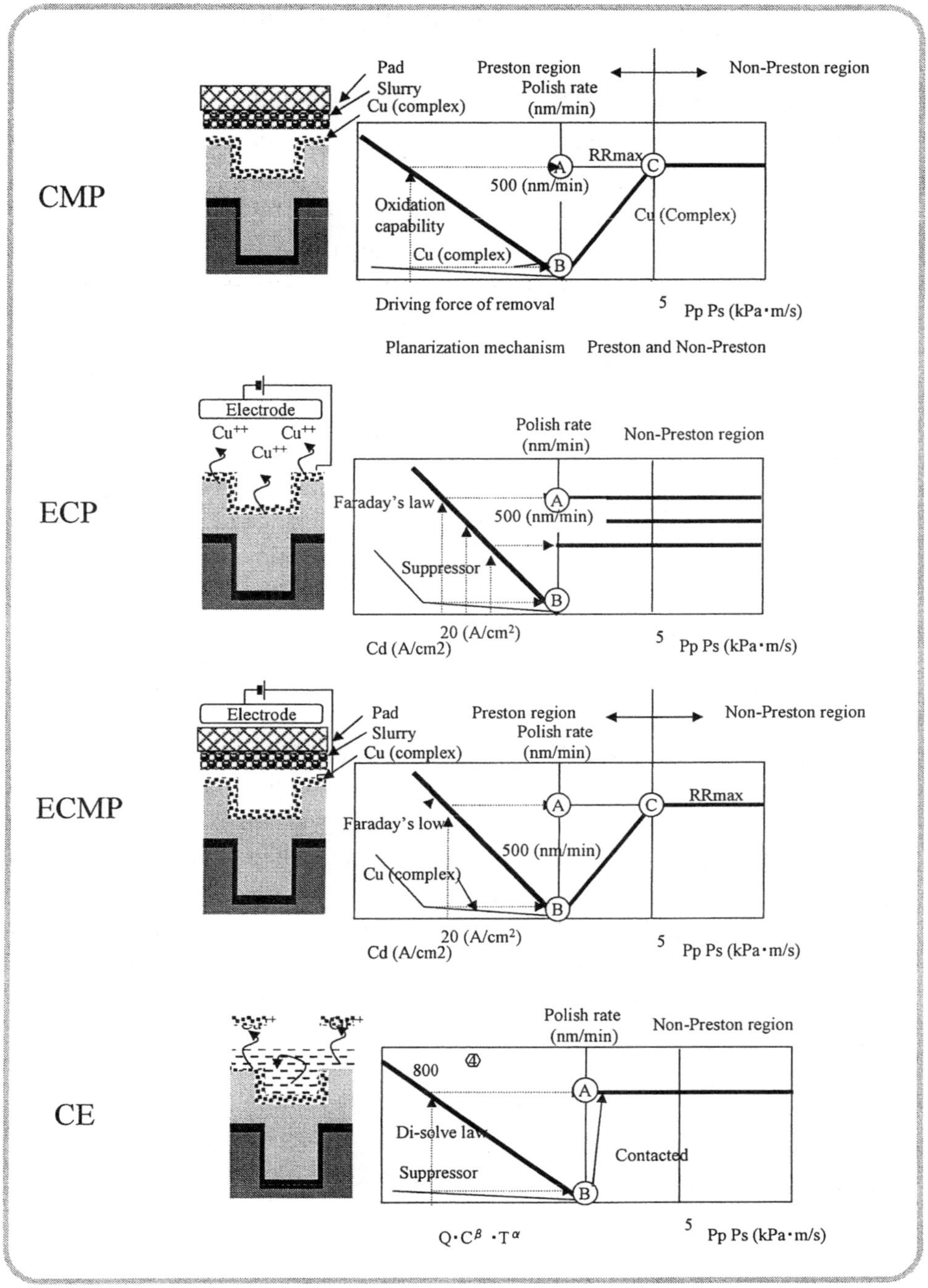
CMP
Pad
Slurry
Cu (complex)
Preston region
Polish rate
(nm/min)
Non-Preston region
RRmax
500 (nm/min)
Oxidation capability
Cu (complex)
Cu (Complex)
Driving force of removal
5
Pp Ps (kPa·m/s)
Planarization mechanism
Preston and Non-Preston
ECP
Electrode
Cu++
Cu++
Cu++
Polish rate
(nm/min)
Non-Preston region
Faraday's law
500 (nm/min)
Suppressor
20 (A/cm2)
Cd (A/cm2)
5
Pp Ps (kPa·m/s)
ECMP
Electrode
Pad
Slurry
Cu (complex)
Preston region
Polish rate
(nm/min)
Non-Preston region
RRmax
Faraday's low
500 (nm/min)
Cu (complex)
20 (A/cm2)
Cd (A/cm2)
5
Pp Ps (kPa·m/s)
CE
Polish rate
(nm/min)
Non-Preston region
800
④
Di-solve law
Suppressor
Contacted
Q·C^β ·T^α
5
Pp Ps (kPa·m/s)

Special theory of general principles of planarization

Special theory of general principles of planarization

The figure on the next page explains planarization technologies and their governing laws in a general manner.

CMP is controlled by Preston's Law that the polishing rate is proportional to a downforce and relative speed. ECP is controlled by Faraday's Law that a polishing rate (volume) is proportional to an electrical charge. CE is controlled by the dissolution law that a polishing/dissolving rate is proportional to a flow rate, concentration, and temperature. The left part of the x-axis shows a driving force for removing, which means the power of oxidation of an oxidizer in CMP (from Cu to Cu^{++}), power of oxidation of electric energy in ECMP/ECP, or power of dissolution in CE. The right part shows a polishing pressure (Pp) and relative speed (Ps) (parameters in Preston's equation) in comparison with polishing. The y-axis shows a polishing rate. The difference in the polishing rate between Points A and B on the y-axis indicates planarization performance.

Comparison with the figure

Let's compare the figure with that on the precedent pages. The comparison facilitates your understanding.

Once you understand these basic technologies, you will also understand their applications. Cu planarization technologies introduced in the current market are expanded versions. Let's think once again. Cu processing technologies in the general field are originally based on etching or ECP. Polishing adopted in the semiconductor market may be rather unusual. Adopting etching and ECP into Cu processing technologies is, therefore, very natural. Such adoption, however, requires elaborations and refinements for planarizing semiconductor devices. Polishing has already been a matured technology, which evolved with various refinements. Etching and ECP matured in the general technological field and polishing technologies in the semiconductor field compete each other for their superiority, but I personally feel that they can co-exist in harmony.

Of course, the conditions of the polished wafer surface greatly affects the result of the subsequent cleaning process, and those of the cleaned surface are influential over the result of the subsequent drying process.

Now, the general theory of general principles of planarization governing the processes including deposition will be discussed from the next section.

(Special) the general principle of planarization

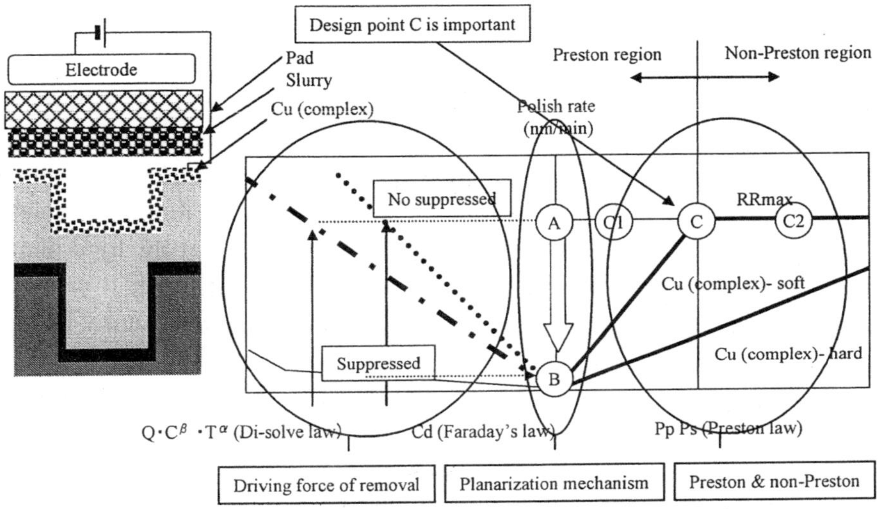

Please compare

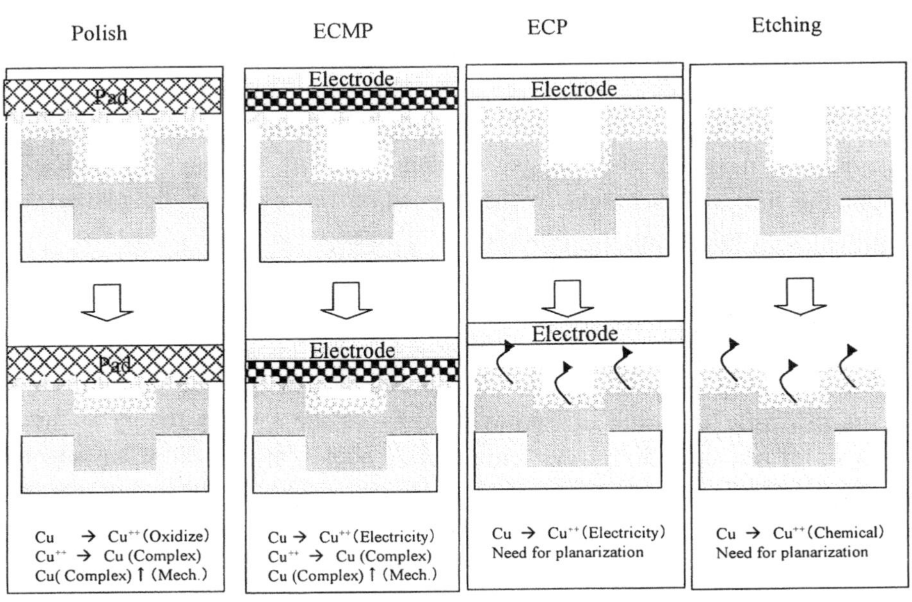

General theory of general principles of planarization

General theory of general principles of planarization, including deposition

First, global and local step heights are explained. As shown in the figure on the next page, Cu deposition by plating involves wafer level non-uniformity, and this non-uniformity is called a global step height caused by plating. At the microchip level, the interconnect density causes a step height, which is called a local step height caused by plating. Through improvements of plating technologies, the global step height has reduced to 5 % (3σ) of the total plating. For the local step height, dishing is the interconnect height itself; it cannot be eliminated by plating but by device designing, such as creating dummy patterns. Mounding is solved by combining device design and plating solution design.

Let's review approaches to planarization realized by deposition and removing. In conventional plating, the presence of dummy patterns determines the dishing size. Plating capable of planarizing the deposition without CMP can minimize the initial step height. The sum of global and local step heights would be the initial step height to be planarized in the subsequent deplating process. If the initial step height is large enough to deteriorate planarization performance in the subsequent removing process, the deposition should be designed to be thicker to fully planarize the plated surface; naturally, CoC would be worse.

The general theory of general principles of planarization is applied to determine the most suitable deposition and removing volumes by taking into account the initial step height in the deposition process and the planarization performance in the removing process. Practically, the input of deposition and removing process data into PC allows us to determine the best planarization process with the least CoC.

I name the principle that focuses on processes from deposition to removing as the *general* theory of general principles of planarization. On the other hand, I call the principle focusing on removing only as the *special* theory of general principles of planarization. As I mentioned above, and without saying, these principles are not academically present. However, I believe that they are useful for people to understand planarization technologies in the damascene process, and my students have affirmed the importance. I truly wish that they also could be of the readers' help.

Let's check for deposition process

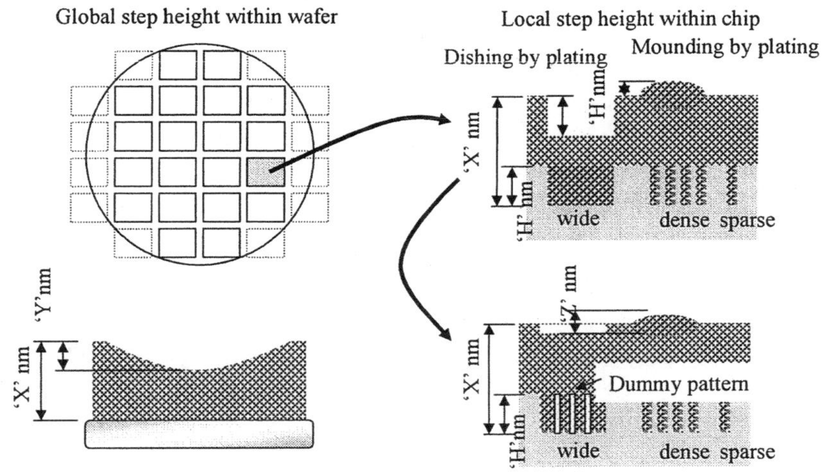

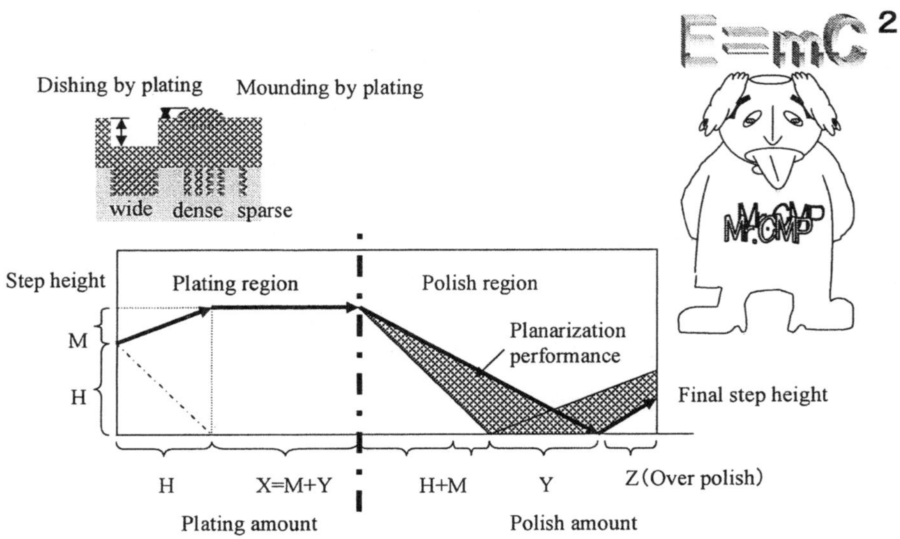

Nanotopography : yield damage of STI

Yields of STI

There is a rumor about STI CMP that wafer waviness may negatively affect the yield after CMP. CMP is basically a process characterized in following large waviness, but not following small waviness in planarization. The degree of waviness can be controlled by the conditions of the pad, slurry, and CMP operation, and such control has been successful in the planarization of semiconductor devices. However, the improvement of the planarization performance of CMP may rather degrade the STI CMP process according to the shape and degree of wafer waviness. This chapter roughly classifies the device film structure into four types in terms of CMP, analyzes STI CMP characteristics by FEM, well known for the analysis of step height performance, and the effect of wafer waviness that exerts STI. Here is the result of these analyses.

Effect of wafer waviness

The wafer shown in the figure on the next page has the waviness with a height of tens of nanometers (peak to valley (PV) value) with a cycle of several millimeters to tens of millimeters (λ). Recently, there has been a controversial report that the waviness, called nanotopography, affects the yield of STI CMP. This waviness is hereinafter defined as nanotopography. For STI, oxide film is first formed according to nanotopography shown in the figure, and then is polished, following the peaks and valleys of the deposition surface in CMP. At that time, the deposition surface may be polished with following the waviness while the peaks of STI may be overpolished without following the waviness. The next section shows the example of analysis about how the nanotopography on wafer affects the STI process from the initial stage through the deposition process to CMP, as well as how it can be prevented.

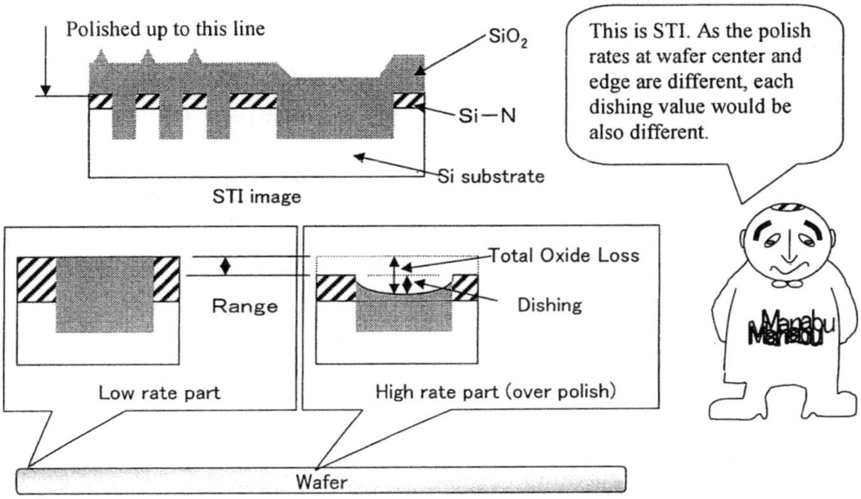
Polished up to this line
SiO_2
Si－N
Si substrate
STI image
This is STI. As the polish rates at wafer center and edge are different, each dishing value would be also different.
Total Oxide Loss
Range
Dishing
Low rate part
High rate part (over polish)
Wafer
Manabu

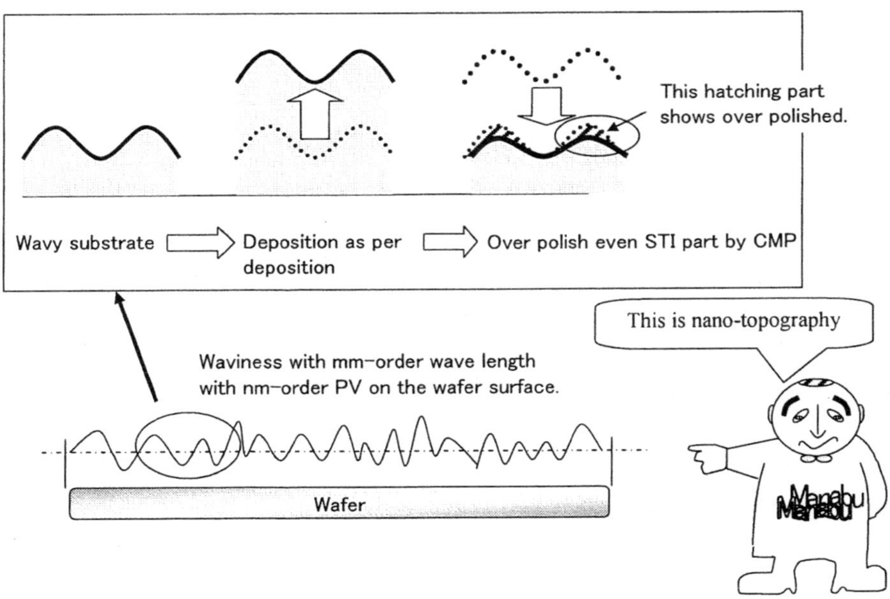
This hatching part shows over polished.
Wavy substrate
Deposition as per deposition
Over polish even STI part by CMP
This is nano-topography
Waviness with mm-order wave length with nm-order PV on the wafer surface.
Wafer
Manabu

Polishing not to follow wafer waviness

Example of effect analysis of wafer waviness

An analysis diagram is shown in the figure on the next page. A condition in which, when the valley is polished with a polishing volume of Vv after deposition, the difference ΔV between the polishing volume of the peak Vp and that of the valley Vv is equal to or less than a certain allowable value is defined as the condition in which nanotopography does not affect STI CMP. Assuming that the slurry selectivity of stopper/STI film is 1, STI is overpolished by ΔV.

$\Delta V = (Vp - Vv) < Val$ at Vv

Where ΔV: Difference in the polishing volume between the peak and the valley (overpolishing volume)

Vp: Polishing volume at the peak

Vv: Polishing volume at the valley

Val: Allowable overpolishing volume

Analysis result and countermeasures

The graph shows a relationship between the cycle λ (characteristic value of nanotopography), PV value, and overpolishing volume ΔV of a device (pad used: IC1000™ + Suba400™, ΔV = 20nm). In the upper left part of the graph, the overpolishing of 20 nm or more is caused because PV is large, and λ is small. In the lower right part, no overpolishing of 20 nm or more is caused because PV is small, and λ i s large. Another graph shows another analysis using pads with different stiffness. A higher stiffness pad decreases the allowable region. This explains that, on nanotopography, safety margin is higher when a pad of lower stiffness, such as Suba™, is used, while the margin is lower when a pad of higher stiffness, such as IC1000™ monolayer, is used.

Countermeasures

Here explains a countermeasure provided by the CMP process. With a slurry selectivity larger than 1, overpolishing due to wafer waviness is decreased with inverse variation to the selectivity. However, the selectivity must not be unreasonably increased because a larger selectivity generally promotes dishing. Typically, the selectivity may be up to about 30. The optimal CMP process must be selected according to the allowable nanotopography of wafers to be used, the range of stopper film thickness, and the allowable value of dishing of semiconductor device design.

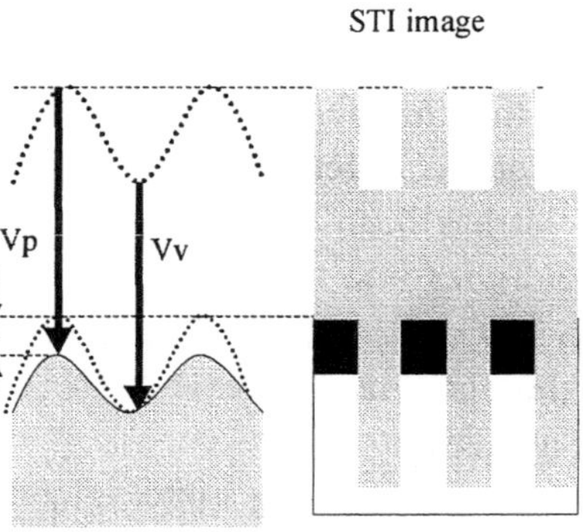

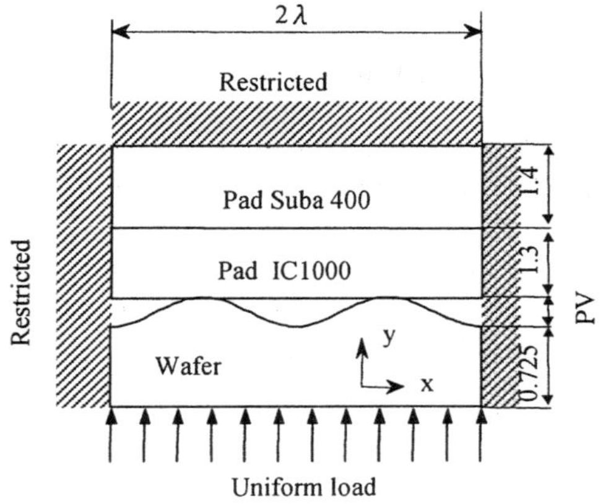

STI on nanotopography

FEM input

- ΔV= 20 nm
- Pad stiffness E= 0.98 MPa

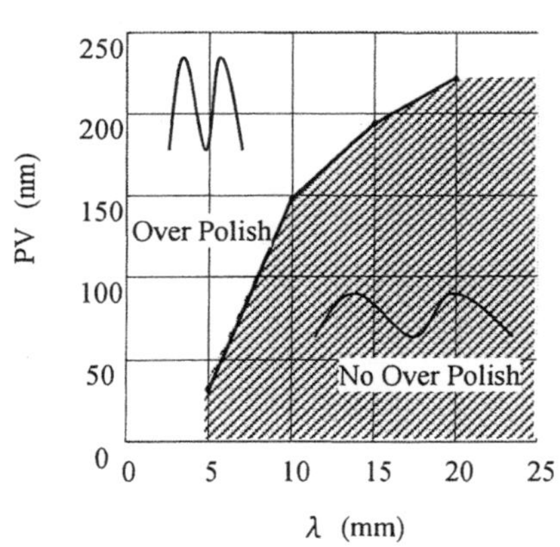

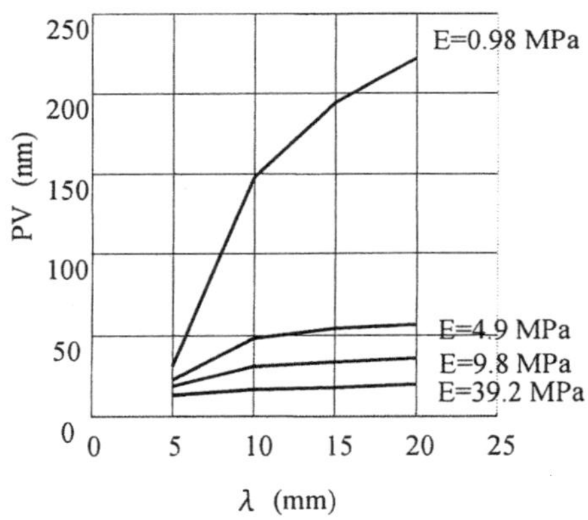

Roll-off: wafer edge hang

Roll-off problem

Typically, edge roll-off (ERO) is seen near the chamfered part of the wafer edge, and its planarization level is lower than at the wafer center; it is slightly more convex than the flat polished surface. It has been known that, even for 300 mm double side polished wafers, ERO intrudes into the fixed quality area (FQA) of the 2 mm EE specified by the ITRS 2003. ERO is now regarded as a concern in the lithography process because the planarization level at the wafer edge required for the latest devices cannot be achieved.

Whether ERO may affect the CMP process or not is a quite interesting question. So, the impact of ERO on the CMP polishing profile is analyzed using FEM.

Analysis method

Supposing that the CMP polishing rate can be expressed by Preston's equation, the polishing profile of a wafer with ERO is estimated by calculating the surface pressure distribution with FEM analysis. For the FEM analysis, only the pressing load in the vertical direction is considered while the rotations of the wafer and the pad are ignored. Also, a two-dimensional cylindrical coordinate system with the axis of symmetry at the wafer center is assumed.

The figure on the next page shows the analysis mesh of the wafer edge. The wafer has a diameter of 300 mm and a thickness of 0.775 mm, and is chamfered to have a trapezoid shape. The polishing pad has a laminated structure comprised of a soft pad (underlying material) and a hard pad (surface layer). The back surface of the wafer is covered with backing film through which a uniformly distributed pressure (wafer pressing pressure) is applied. The wafer circumference is pressed with a retainer ring (RR).

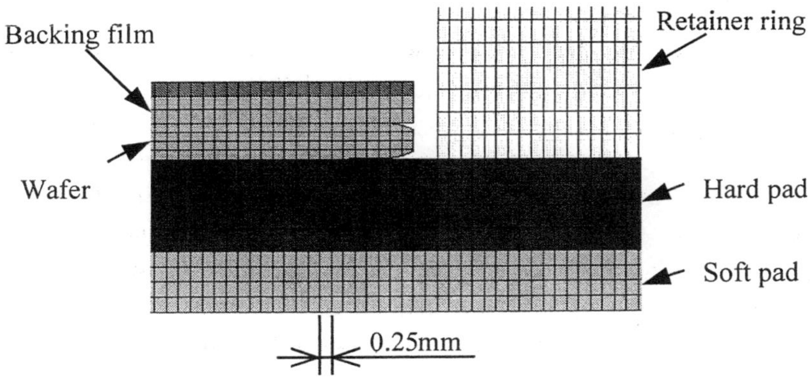

FEM mesh

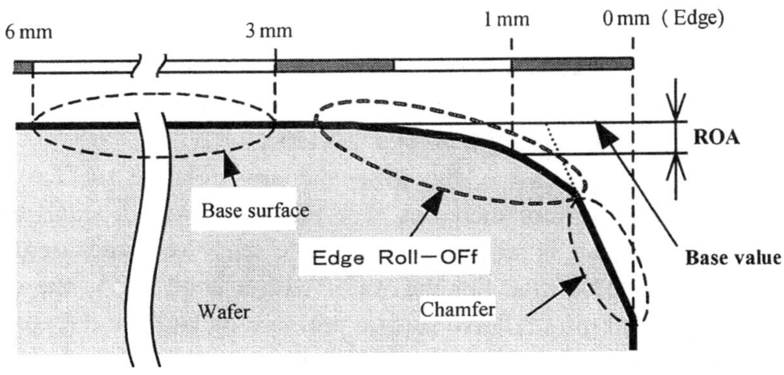

Definition of ROA

Targeted edge: 1 mm

Impact of ERO

The figure on the next page shows the surface pressure distribution at the wafer edge with both wafer pressing pressure and RR pressure set to 30 kPa. The y-axis shows the relative surface pressure, supposing the surface pressure at the wafer center to be 1. As described above, the surface pressure curve almost agrees with the polishing profile curve. For the wafer with a small roll-off amount (ROA), the relative surface pressure is almost uniform inside a radius of 148 mm. For the wafer with a large ROA, the relative surface pressure increases outside a radius of 146 mm. It is found that the surface pressure at a 148 mm radius point is higher by about 30 % than at the wafer center.

If the allowable variation of the polishing amount in FQA with 2 mm EE is ± 10 % (relative surface pressure of 1.0 ± 0.1), the relative surface pressure is within the allowable range for the wafer with a small ROA, but it is largely out of the range for that with a large ROA. The result shows that ERO greatly impacts the polishing profile at the wafer edge and is an obvious problem for CMP.

Impact of RR pressure

The graph shows a relationship between the RR pressure and the max. and min. relative surface pressures in FQA with 2 mm EE for different wafers. The wafer pressing pressure is fixed at 30 kPa this time.

In the range of RR pressure, the max. and min. relative surface pressures increase as the RR pressure increases. For the wafer with a small ROA, the relative surface pressure is within the allowable range (colored area) at a RR pressure of 22.3 to 23.8 kPa. For the wafer with a large ROA, the difference between the max. and min. relative surface pressures is large, and there is no RR pressure that corresponds to the allowable range.

This result suggests that, if ROA is small, ERO can be addressed by adjusting the RR pressure. If ERO is excessive, however, the allowable range is exceeded even when the RR pressure is adjusted.

The impact of ERO on the CMP polishing profile has been analyzed with FEM. The result shows that ERO greatly impacts the polishing profile at the wafer edge; if ROA is 1.27, the surface pressure at a 148 mm radius point is higher by about 30 % than at the wafer center. However, ERO can be addressed if its level is allowable for CMP; it can be overcome by adjusting the RR pressure if ROA and its variation are small.

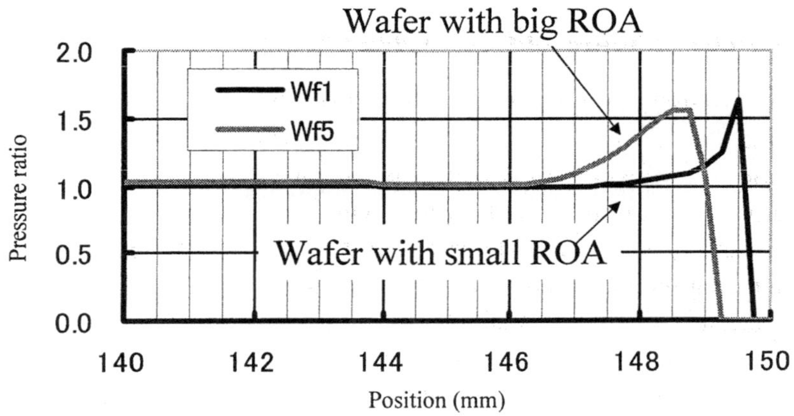

Polish profile

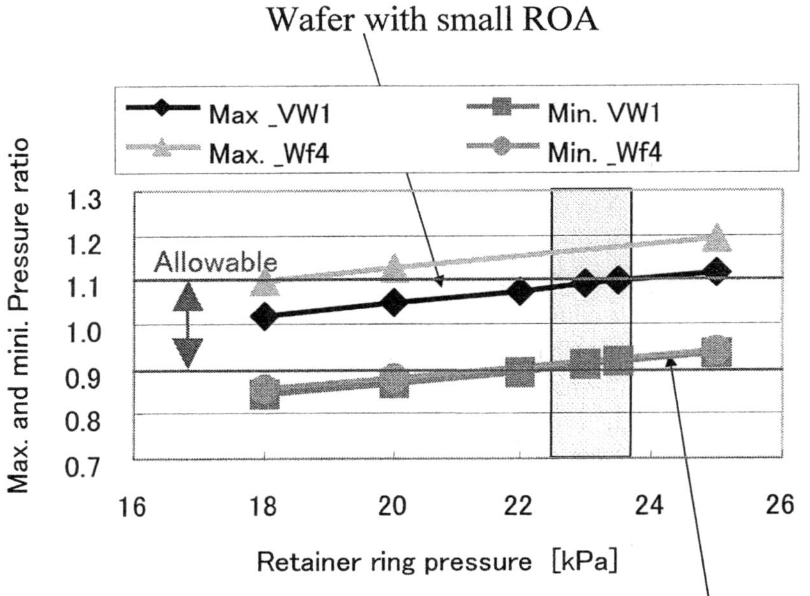

Effect by retainer ring pressure

FEM effective from km to nm

Structure analysis applicable at the nano level

As the integration and the speed of LSI increase, the use of low-k material with a dielectric constant lower than that of conventional SiO_2 is under consideration. Promising low-k materials include porous materials with many pores (relative permittivity: 1) and organic materials. These materials, however, have low mechanical strength; devices may be damaged in the manufacturing process, reducing the yield and reliability. In the semiconductor manufacturing process, various kinds of load (heat, process reaction, etc.) affect the interconnect structure. This section estimates the mechanical characteristics of materials that will be adopted in the future and analyzes stress that will affect the interconnect structure during CMP with FEM.

Analysis conditions

Stress generated during CMP to an interconnect structure formed by a dual damascene method is analyzed with FEM assuming the mechanical characteristics of future dielectric materials, interconnect size, and structure.

1) Interconnect structure/size: The figure on the next page shows the schematic diagram of a via and interconnect layers on it. The table shows the assumed sizes with the via diameter set as the representative size in nm. The analysis is conducted for three structures: (1) a homogeneous structure without HM (E1, E2, and E3 are made of the same material), (2) a homogeneous structure with HM (the material of E3 is different from those of the others), and (3) a hybrid structure with HM (the materials of E1 to E3 are different).

2) Mechanical characteristics of materials: Young's modulus of E1 to E3 are estimated based on bulk relative permittivity proposed by the ITRS for each technology node (TN). The interconnect material (Cu), barrier film (Ta/TaN), and etch stopper (SiCN) are fixed regardless of the size.

3) Analysis mesh/load condition: The analysis uses a 3D model corresponding to 1.5 layers, including a simulated maximum width interconnect beneath an interconnect (minimum width) and via, as shown in the figure. At M1 = 100 nm, the size of the model is 1•m x 1•m x 0.6•m. It is assumed that the model's bottom layer is bound with the axis of symmetry at the via center. At both ends of the analysis mesh, semi-infinite elements represent the infinite direction. For CMP, a vertical load of 3.45 kPa (0.5 psi) and a shear stress with a friction coefficient of 0.43 are evenly applied to the entire top surface of the model.

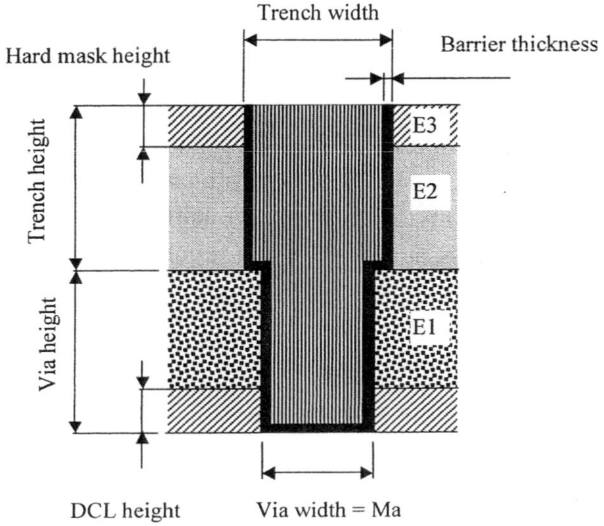

Damascene structure

FEM input values (nm)

M1		140	100	70	50
Trench	W (min)	160	114	80	57
	W (max)100	0	500	300	100
	Height	230	170	125	90
Via	Dia.	140	100	70	50
	Height	210	150	110	80
Hard mask	Height	50	35	25	25
DCL	Height	50	35	25	20
Ta/TaN	Thickness	10	7	5	3.5

Confirm strength of the damascene structure

Is the root subject to the highest stress?

The figure on the next page shows the distribution of von Mises equivalent stress viewed from the axis of symmetry at the via center in the interconnect structure, as well as an example of deformation due to stress. The via bottom is supported by an etch stopper (SiCN); the middle part or the via root is subject to the highest stress. This analysis result is reasonably convincing.

The graph plots the stress at the via as the function of the scaling generation M1. It will help you understand how the stress varies with the generations.

Countermeasures

Predicting the evolution of low-k material along with the improvement of LSI performance, stress applied to the interconnect structure during CMP is analyzed. Stress at the via tends to depend on Young's modulus of low-k material, but it will not significantly change. This suggests, therefore, that the stress problem is unlikely to be serious as it is currently concerned in the future when the 32 nm generation is achieved in 2010. Regarding stress at the via, it may be unnecessary to extremely reduce the processing pressure for CMP.

In this analysis, the von Mises equivalent stress has been examined with respect to deformation and breakdown. The analysis result explains that FEM is useful for analysis both at the kilometer level (mechanical engineering) and at the nanometer level (semiconductor processing). It is recommended for all engineers to maximize the use of FEM.

Lastly, I would like to answer the question if you ask whether the stress analysis is inadequate for fragile low-k materials. The answer is yes. Indeed, analysis based on fracture stress is required to analyze stress affecting the interconnect structure in terms of delamination and shear. However, the stress level to be used for such analysis is obtained in the same way as described above, that is, by using FEM. The difference is whether the obtained stress is considered as allowable stress for analysis or it is recalculated to determine energy as fracture stress. In addition, fracture stress can be analyzed in the same way as above.

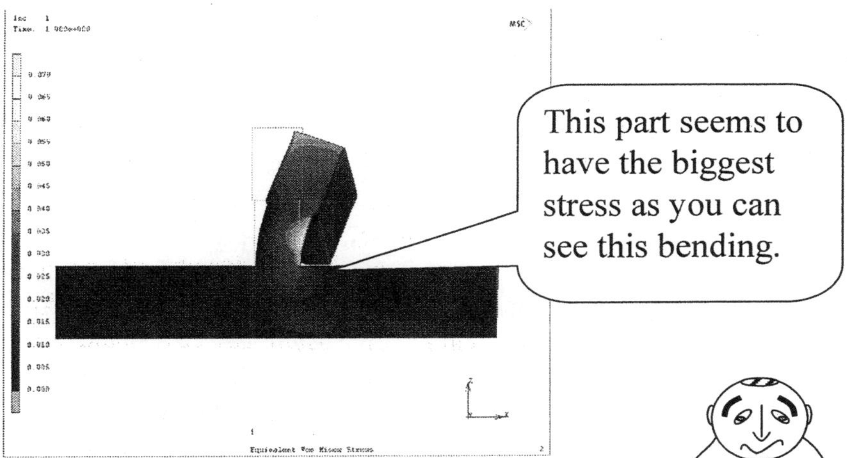
This part seems to have the biggest stress as you can see this bending.

FEM output

Manabu

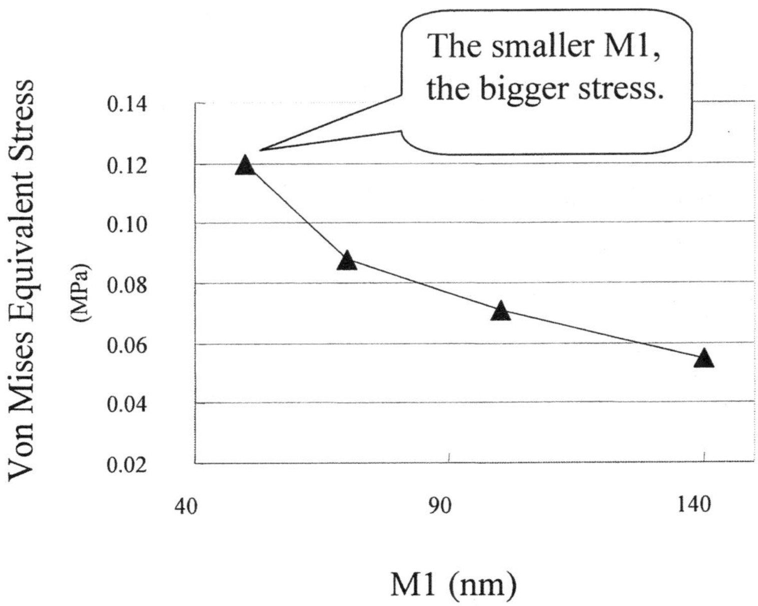
The smaller M1, the bigger stress.
Von Mises Equivalent Stress
(MPa)
0.14
0.12
0.10
0.08
0.06
0.04
0.02
40
90
140
M1 (nm)

= Who's law? =

For the last 60 years, the semiconductor industry has enjoyed success through repeated scaling and wafer size transition. Who has driven the evolution of semiconductor technology, then?

Moore's Law (an empirical rule), formulated by Dr. Gordon Moore, co-founder of Intel, is famous in terms of scaling. This Law is theoretically supported by the scaling theory for CMOS devices formulated by Dr. Robert H. Dennard, an IBM Fellow. Based on Intel's device production technology, Dr. Moore *empirically* predicted that the integration density of semiconductor devices would double every 18 to 24 months. Dr. Dennard *academically* verified in his published paper that the scaling of CMOS devices would provide (1) higher speed, (2) lower power consumption, and (3) lower cost. Since then, the scaling of semiconductor devices has continued with DRAM technology as a driving force. Regarding memory technology, Hwang's Law (Samsung's mission of doubling density growth), stating that memory density would double every year, is also well known.

A question may be raised as to whether there is any rule for wafer size transition. When the number of plants for 8-inch wafers reached 300, the wafer size shifted to 12 inch. Applying this cycle, it may be predicted that the wafer size should shift to 18 inch when the number of plants for 12-inch wafers reaches 300. Who will verify and establish this possible rule?

The great paradigm shift in the semiconductor industry, including three-dimensional (3D) implementation and emerging research devices, will continue in the future. Who will lead this paradigm shift?

Who's law? The semiconductor world is in need of the appearance of heroes.

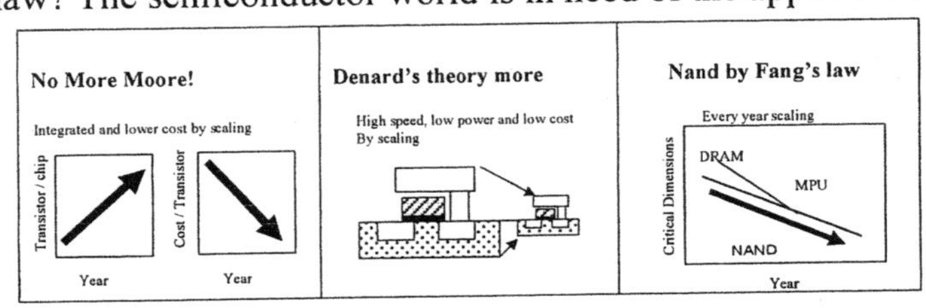

Chapter IV

Wet Surface Treatment : Cleaning and Drying

Cleaning battle

Cleaning of semiconductor devices

The most serious mistake in washing or cleaning may be washing clothes of which color is easy to fade together with light-colored clothes or making your sons and daughters with *dirty feet* help cleaning. All of such problems correspond to *contamination* in the field of wafer cleaning technology.

Various materials exist at the same time on the cleaning surface of semiconductor wafers. To clean the surface with Cu interconnects exposed, Cu on the neighboring oxide film is removed. At that time, Cu interconnects are also swept with the pad; this is like trying to wipe off ink, sweeping the ink bottle. Such a method cannot completely clean the wafer surface!

First of all, the basics of wafer cleaning must be learned.

Non-contact cleaning and soft cleaning have recently been developed. The former is designed to avoid contacting Cu interconnects, and the latter is designed to prevent damage to Cu interconnects. Drying is also a troublesome process. A drying method that leaves no drying spot is essential. Metaphorically speaking, it is equivalent to finishing with no scent fabric softener.

History of wafer cleaning

The figure on the next page outlines the history of wafer cleaning. In the past, the batch RCA cleaning process developed by RCA Corporation was dominantly used. After wet process systems for CMP and plating emerged, the dry-in/dry-out concept became a de facto standard, and the idea of incorporating a built-in single wafer cleaning subsystem into each system was generally recognized. Since then, various types of cleaning systems and processes suitable for single wafer cleaning have been developed. At present, *one-bath type* [3)] systems that perform all processes in one bath and *vertical* systems that clean wafers in a vertical position are available. Also, processes using functional waters, such as ionized water and ozonized water, are also available. The development of new cleaning technology will necessarily become a key in the generation of new materials and/or high aspect ratios specified by the ITRS.

Cleaning can be divided into numerous different processes and technologies. The wafer cleaning industry is in a *battle* now.

Do you know the basic rule of cleaning?

Battle of cleaning

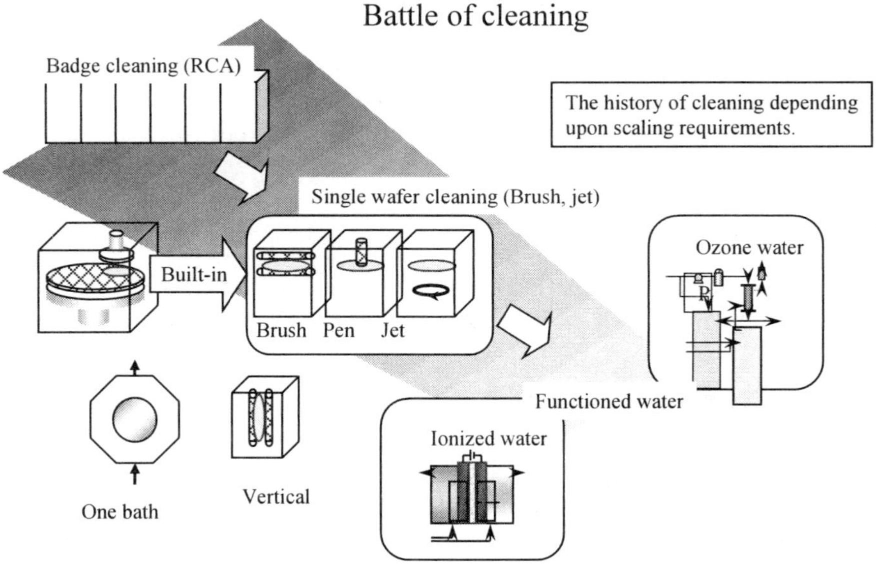

How and what to clean?

This section explains contaminants to be removed and mechanisms for their removal with respect to CMP and plating.

Contaminants to be removed

First, let's look at CMP. Particle and metal contaminants exist on the wafer surface after the CMP process, such as residues of slurry selected for each process (silica, alumina, manganese, etc.), pad debris, and polished debris. They must be removed by cleaning the top, back, and side surfaces of wafers without promoting microscratching developed during polishing. The cleaning region includes oxide film, tungsten, interconnect materials, and barrier materials, which are subject to CMP as described above, and the suitable cleaning method depends on the combination of contaminant types and cleaning regions. Especially in case of using chemicals, the cleaning method cannot be selected by the CMP manufacturer only since the cleaning performance and the availability of subsequent device processes may be limited.

The figure on the next page shows wafer surface materials to be cleaned after the CMP process. Different surface materials are dealt with in different processes. Note that, for example, Cu interconnects have surface materials other than Cu; there exist various related materials on the surface, including Si, SiO_2, Si-N, W, Ti, Ti-N, Cu, Ta, and T-N. Pay attention to this fact when considering countermeasures against zeta potential and corrosion. For plating, impurities contained in the plating solution, such as metal ions and organic substances, and dust generated from the system must be considered.

Cleaning mechanism

The figure shows schematic diagrams of a contamination situation and a cleaning mechanism for CMP. CMP may involve contamination by abrasives, metals, and organic substances (TOC) generated from the slurry, pad, and system. The basic cleaning mechanism may be divided into four steps below.

- Mechanically remove contaminants from the cleaning surface using a brush or jet flow.
- Control zeta potential by adjusting the pH level of cleaning solution in media so that electrical coupling between contaminants and the cleaning surface becomes weak.
- Slightly etch off materials to be cleaned together with contaminants.
- Dissolve contaminants/remove them with the cleaning solution's chelate effect.

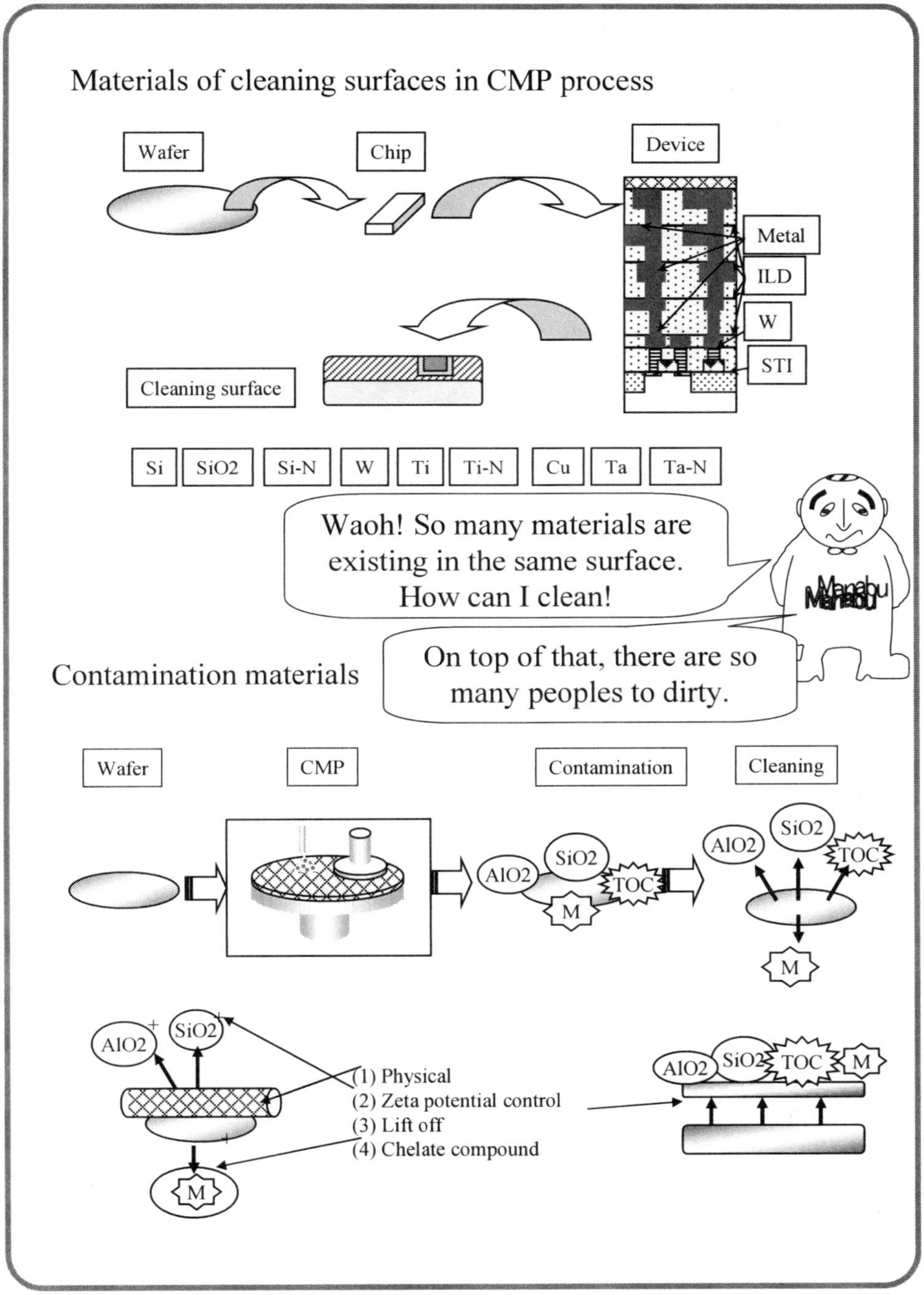
Materials of cleaning surfaces in CMP process
Wafer
Chip
Device
Metal
ILD
W
STI
Cleaning surface
Si
SiO2
Si-N
W
Ti
Ti-N
Cu
Ta
Ta-N
Waoh! So many materials are existing in the same surface. How can I clean!
Manabu
On top of that, there are so many peoples to dirty.
Contamination materials
Wafer
CMP
Contamination
Cleaning
AlO2
SiO2
TOC
M
(1) Physical
(2) Zeta potential control
(3) Lift off
(4) Chelate compound

Start cleaning with clean tools

The figure on the next page shows examples of cleaning modules. Their features are as below.

Roll type cleaning

This is the basic type of single wafer cleaning that cleans the top and back surfaces of wafers at the same time by rotating roll members (PVA, etc.) with several hundred rpm. The roll member shape is based on the accumulated know-how of system manufacturers. There are various cleaning recipes; some control zeta potential by adding alkali solution of about 100 ppm while others add chelate solution to facilitate Cu complex removal. This type must support the use of acid and alkali chemicals. Introducing a small amount of chemicals may improve the particle removal effect, extend component life, and facilitate metal contaminant removal.

Pen type cleaning

This type cleans wafer surfaces by rotating pen-shaped members (PVA, etc.) with several hundred rpm in the same manner as the roll type cleaning. It is called pen type cleaning since it uses pen-shaped (not roll-shaped) cleaning members. It requires the same functions as the roll type cleaning to introduce chemicals.

Jet type cleaning

Three types are available: basic jet, ultrasonic jet, and cavitation jet. The cavitation jet type can be used in a wider range of cleaning with higher efficiency than the conventional type. Two-fluid jet cleaning using two different fluids (gas and liquid) has also been developed. Introducing chemicals may be possible.

Drying

Single wafer drying typically uses the spin rinse dry (SRD) method that rotates wafers with several thousand rpm for final drying. SRD offers good performance, including throughput, for the surface of hydrophilic film, such as oxide film. For the surface of hydrophobic film, such as low-k material, a drying method using isopropyl alcohol (IPA) to prevent watermarks is also under development. These methods may be selected according to the type of film surface.

The cleaning/drying modules described above offer optimum cleaning/drying performance in systems well controlled based on the dry-in/dry-out concept.

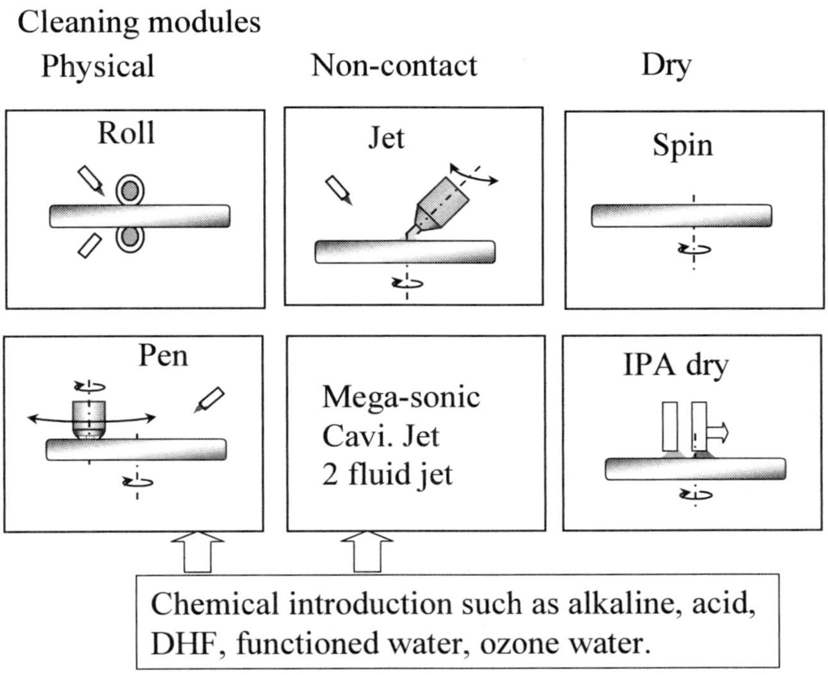

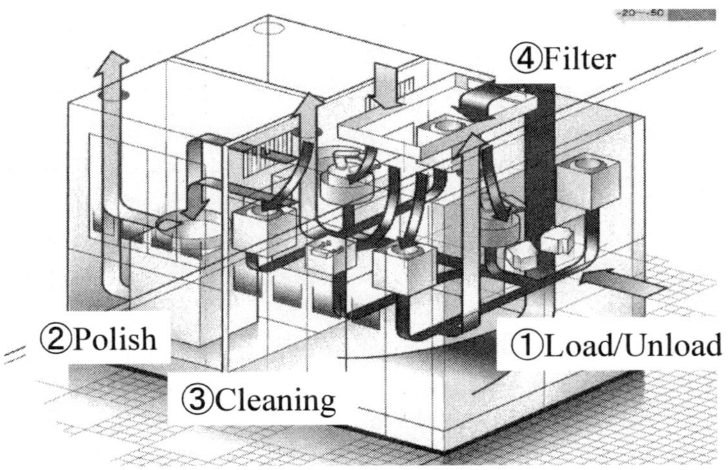

Clean CMP concept

Clean environment

Protection of the clean environment (waste treatment for new materials)

Countermeasures against environmental contamination outside a clean room are discussed below. A wide range of new materials is used for semiconductor devices. The interconnect material has been changed from Al to Cu and will be further changed to alloys. For ILD, the development of inorganic and organic low-k materials is booming. Candidate capacitor materials include high-k materials (Ta_2O_5, STO, BST, etc.) and noble metals (Pt, Ru, Ir, etc.). As the transistor gate material, high-k materials, such as HfO_2, have been adopted. Various semiconductor device materials are available now. It takes time even for me to find these elements in the periodic table of the elements. Processing devices made of these materials naturally produces waste containg the materials or their by-products, making wastewater treatment more complex. Wastewater treatment is not only the issue for CMP, but it is desirable to start with CMP, which is known as one of the most polluting processes (this is nothing to brag about). CMP has actually been developed with environmental consciousness from the beginning.

For example, the recycling/reuse of slurry has been attempted, and wastewater is reused in closed systems. For polishing bare silicon wafers, slurry is reused several times. This is possible since the polishing of oxide film or polysilicon in semiconductor device processing does not generate by-products containing different materials. On the other hand, wastewater recycling has disadvantages; whether to adopt or not depends on the choice of engineers. Wastewater treatment for CMP is same as that for other processes except that wastewater from CMP contains abrasives. Wastewater treatment system manufacturers can develop systems for CMP just like required for other wastewater treatment. Such systems are easy to adopt; whether to adopt or not depends on the choice of engineers.

If the use of new materials as described above increases in the future, the compositions of slurry and wastewater will change. It is desirable to develop CMP processes considering the selection of environmentally friendly slurry and wastewater treatment from the beginning, instead of facing the *requirement of wastewater treatment* after completing development.

Countermeasures against environmental contamination are essential and should be taken to eliminate the shameful recognition of CMP as *one of the most contamination processes*.

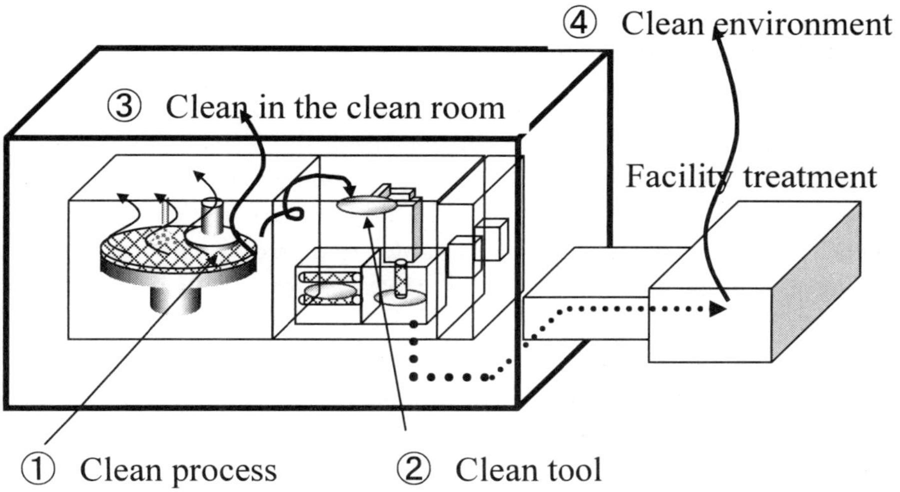

Several clean technologies

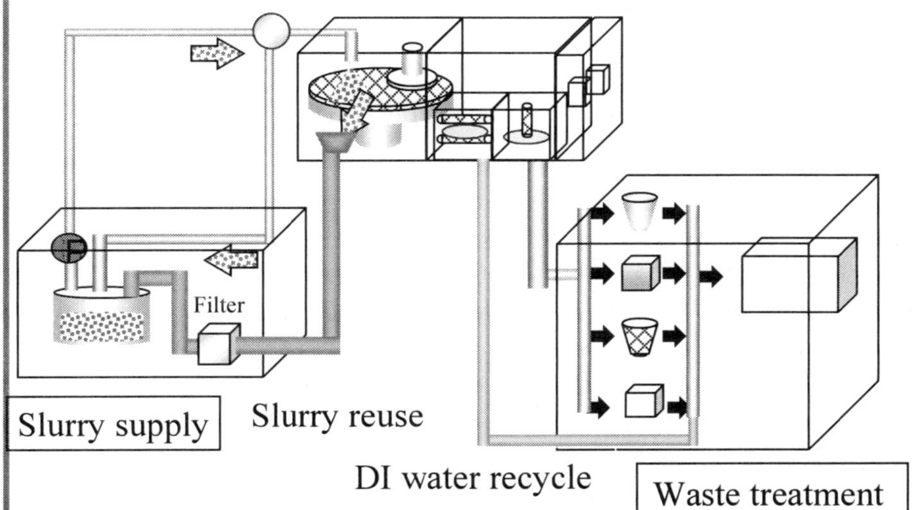

Clean environment

Non-contact cleaning

Demands for non-contact cleaning

Stricter particle control is required as scaling proceeds. In order to control particles at the size of half the critical dimension (CD), particles having the diameter of 22.5 nm must be detected for the 45 nm generation. Such size is almost equal to that of dishing formed in CMP; it is becoming more difficult to distinguish whether a detected defect is a scratch or a particle only from the size. Accordingly, the cleaning approach is shifting from contact cleaning to non-contact cleaning.

The most common type of non-contact cleaning is based on water jet. In addition, other various types of non-contact cleaning are under development, such as water jet cleaning using ultrasonic waves, ice jet cleaning, and jet cleaning with two different fluids (liquid and gas). This section introduces a cavitation-based cleaning method as an example of non-contact cleaning.

Cavitation jet is a type of non-contact cleaning like water jet. It uses ultrasonic waves and is easy to change functional water to be injected according to the purpose. It can also provide a high degree of pressure amplification and a wide pressure receiving area.

Principle of cavitation jet

Cavitation jet cleaning is a wafer cleaning method based on the physical force of cavitation, which is often considered as unfavorable in the field of fluid engineering. If the pressure of fluid falls below the saturation vapor pressure, cavities are generated in the fluid. These cavities collapse as the fluid pressure rises to or above the saturation vapor pressure, producing high pressures at the instant of collapse.

For the test system shown in the figure on the next page, high-speed jet water is flown in the midst of low-speed jet water. The difference in speed between the high-speed flow and the low-speed flow generates vortexes in the boundary region. A low-pressure area is formed at the center of each vortex, producing a cavity. Cavities collapse on the wafer surface and produce cavitation collapse pressure to effectively remove particles on the wafer surface.

What is cavitation?

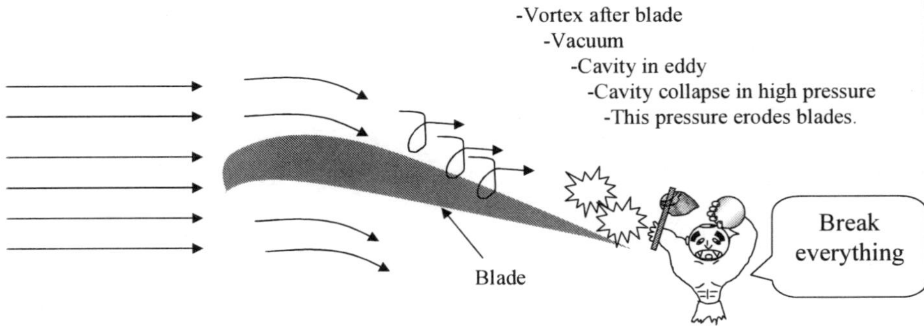

Cavitation should be avoided in fluid dynamics.

Mechanism of cavitation jet

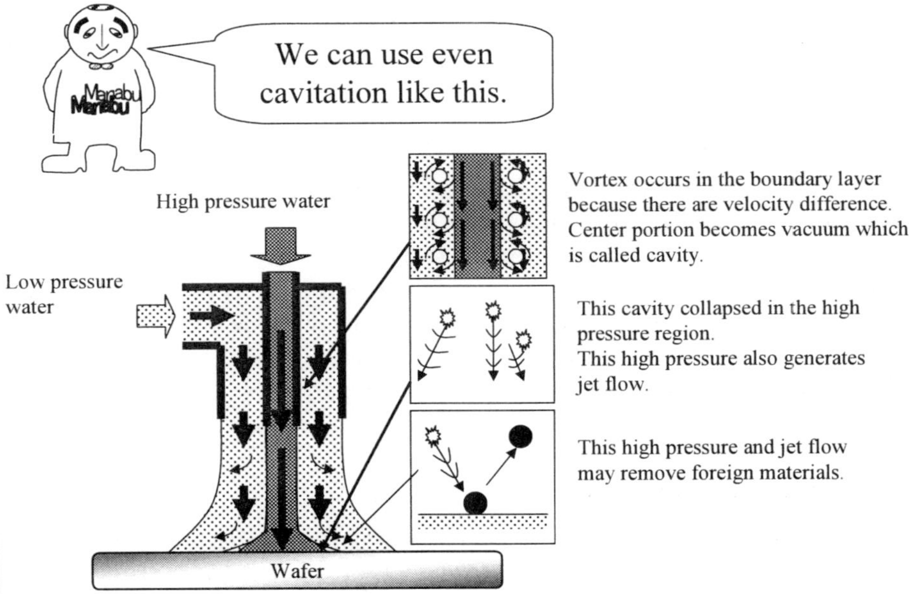

Tame the wooly horse!

Observation of cavitation

Cavitation conditions were directly observed using a high-speed video camcorder. During the observation, the pressure was changed in a range between 1 to 13 MPa. It was like the observation of a storm. Indeed, the use of cavitation jet was like an attempt to ride a wildly bucking horse.

Wide pressure receiving area

The pressure receiving area was compared between cavitation jet and water jet by using the standoff distance as a parameter. The result confirmed that the pressure receiving area of cavitation jet was several tens of times wider than that of water jet. A wider pressure receiving area can reduce the time of single wafer cleaning as if a larger broom cleans up quicker.

Mixture of frequencies for diverse cleaning needs!

Cavitation jet was generated at several frequencies up to 60 kHz. Compared with conventional ultrasonic cleaning using a single frequency, cavitation jet was confirmed to support a wider range of frequencies, offering a suitable frequency for each wafer size.

Several times higher pressure!

The cavitation jet pressure on the wafer surface was compared with the water jet pressure by using the standoff distance as a parameter, as shown in the middle right graph of the figure on the next page. The result showed that the pressure amplification ratio of the former was about twice as high as that of the latter. In addition to the cavitation collapse pressure at the local level, cavitation jet is expected to provide a pressure about twice as high as that of water jet at the global level. Know-how is required for controlling the pressure. Would it be an exaggeration to say an excess cavitation jet pressure can pierce a rock?

High cleaning performance!

Cleaning performance was compared between water jet and cavitation jet based on the slurry residues (%) against the initial slurry amount (100 %). For the former, the slurry residues were about 20 %; for the latter, they were 10 %. The time elapsed until the slurry residues were decreased to 20 % was 30 seconds for water jet and 12 seconds for cavitation jet. This result shows that cavitation jet is superior to water jet in terms of cleaning performance/time.

Cavitation test

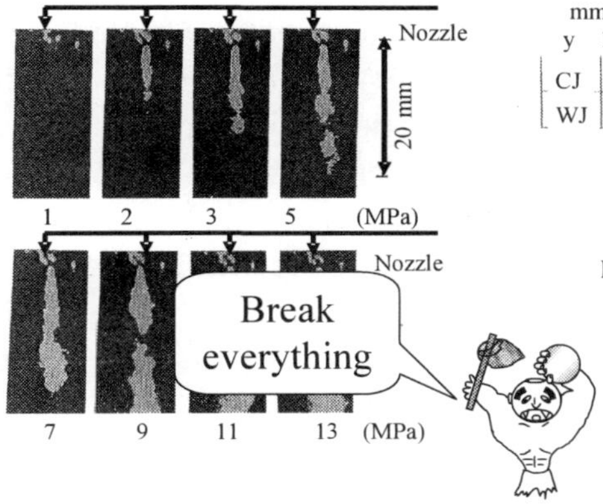

Wide impinging area

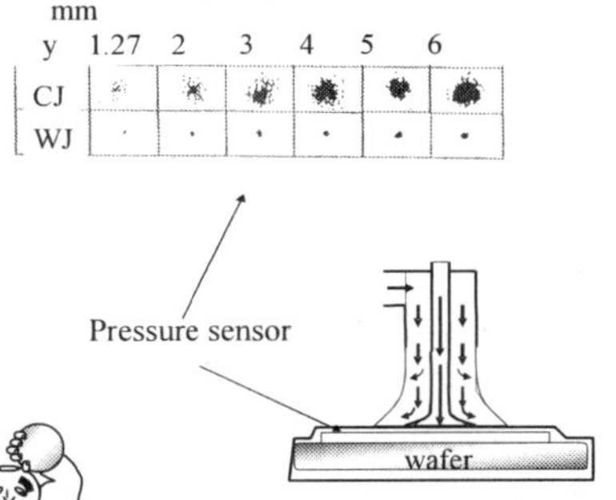

Existing several frequencies may be effective.

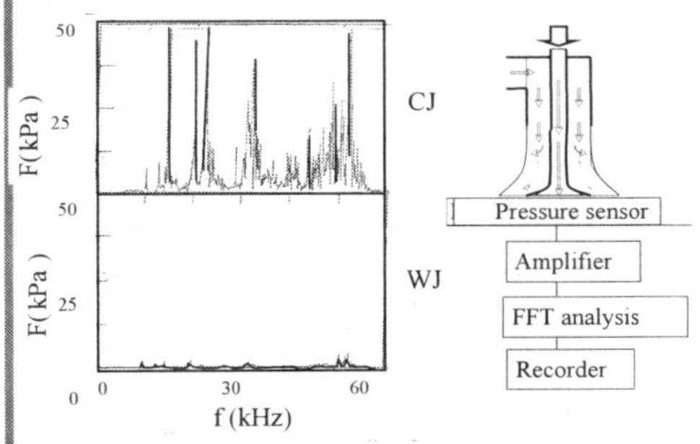

Pressure becomes several times.

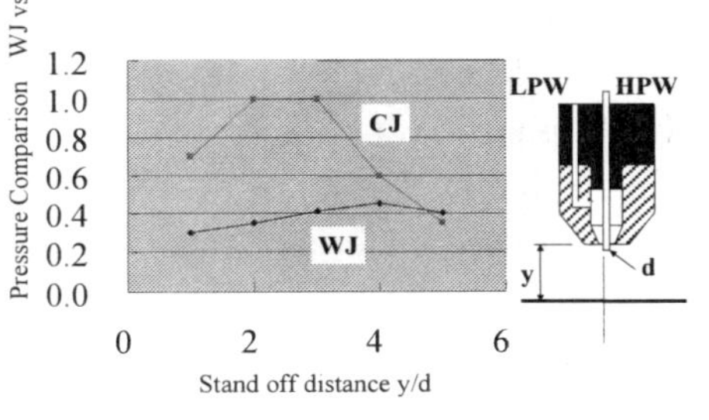

Better cleaning performance

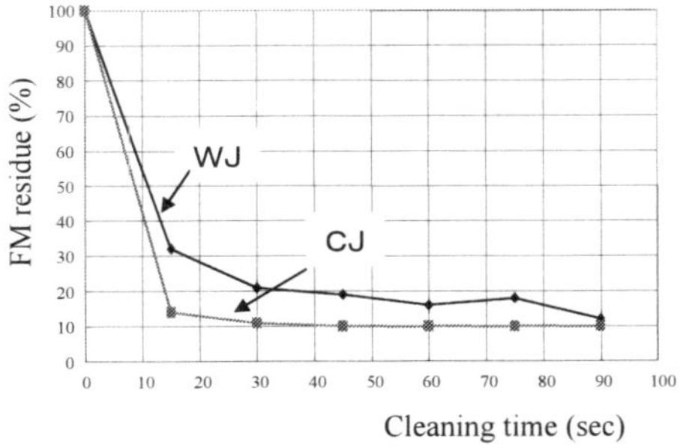

Tool image

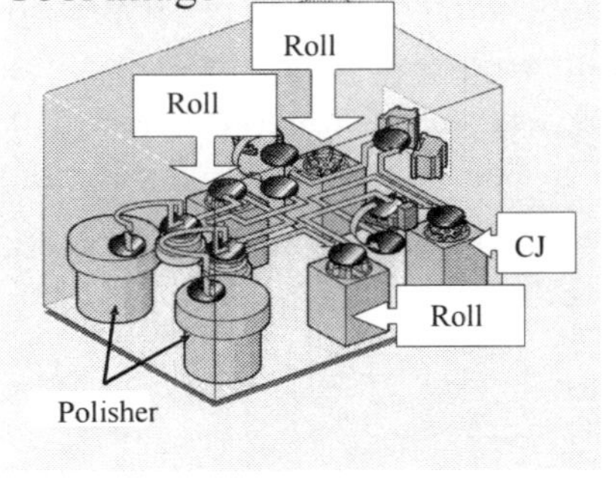

IPA for hydrophobic surfaces

What is IPA drying?

The CMP process for Cu/low-k interconnects requires polishing, cleaning, and drying technologies that cause no damage to low-k film. The surface of hydrophobic low-k film tends to generate watermarks, which have been reported to affect electrical characteristics. IPA drying technology was first invented to prevent watermark formation on hydrophobic film in the front-end processes. As described above, this technology has been recently used for the cleaning and drying of hydrophobic low-k film in the interconnect process. There are also cases of applying IPA drying to hydrophilic film. However, it will be required to verify whether IPA drying is effective for hydrophilic film.

IPA drying includes Marangoni drying (submerging wafers in IPA) and Rotagoni™ drying (developed by IMEC for the face-up process). This section describes drying technologies for CMP. The effect of Rotagoni™ drying on hydrophilic and hydrophobic films on the wafer surface is compared with that of common SRD below.

Hydrophobic film comprises a substrate that is difficult to dry. However, the drying of hydrophobic film is required to achieve the drying of the latest devices based on low-k material. This is a big challenge.

Rotagoni™ drying

The top figure on the next page shows a schematic diagram of a Rotagoni™ drying unit. The unit supplies IPA vapor and D.I. water from nozzles above the rotating wafer while moving the nozzles from the wafer center to the wafer circumference. Rotagoni™ drying is based on the centrifugal force of rotating wafer and the Marangoni effect using IPA vapor. A schematic diagram of a CMP system incorporating the Rotagoni™ drying unit above is also shown.

Rotagoni™ type IPA drying module

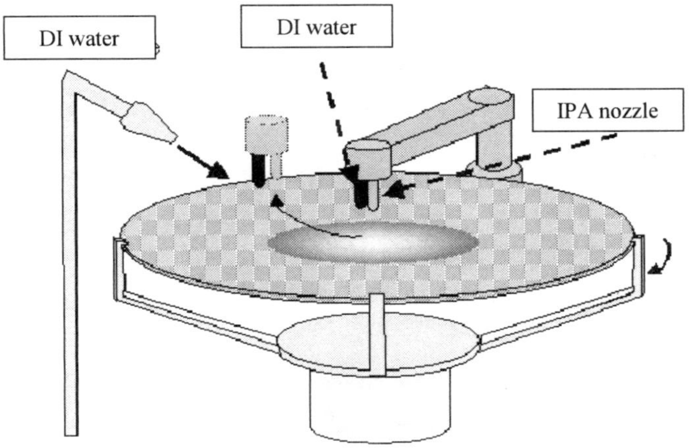

CMP with Rotagoni™ type drying module

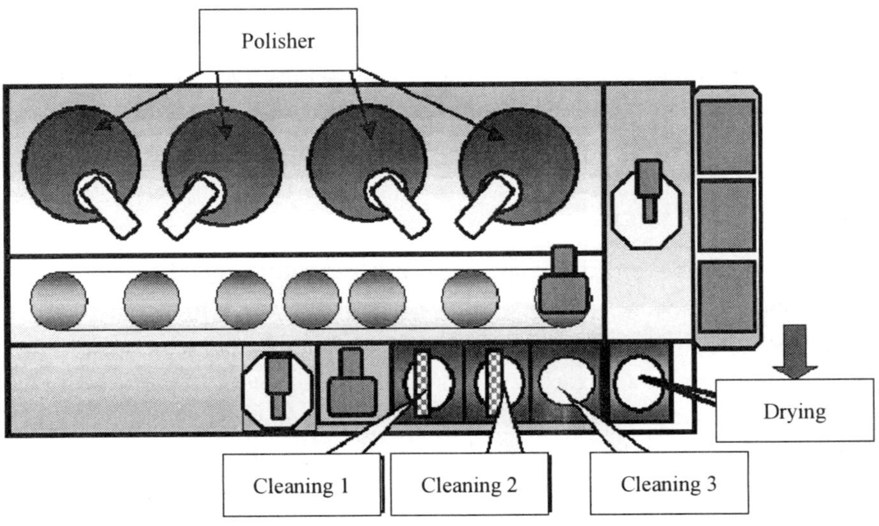

Wonderful! Easy hydrophobic film drying

The effect of IPA drying verified

Various types of film, such as hydrophilic oxide film and hydrophobic polysilicon, comprise a substrate. Furthermore, the hydrophobic surface may be modified to have a hydrophilic property after CMP. IPA drying is effective for hydrophobic film while SRD is desired for hydrophilic film in terms of throughput.

The effects of SRD and IPA drying are compared by changing the hydrophobic (hydrophilic) level through the surface treatment of bare silicon wafers.

Drying performance

The graph on the next page shows the number of defects at each contact angle on the wafer surface after SRD and IPA drying of bare silicon wafers. Generally, the contact angle of a droplet indicates the hydrophobic/hydrophilic level on the wafer surface. For hydrophobic film with a contact angle of about 60 degrees, the number of defects was smaller in any IPA drying than in SRD. For hydrophilic film with a contact angle of about 30 degrees, the number of defects was almost equal between SRD and IPA drying.

This result demonstrates that, for hydrophobic film, IPA drying is effective for reducing the number of defects. It also shows that, for hydrophilic film, there is no difference between SRD and IPA drying (no dependency on the drying method is observed).

SRD and IPA drying

As described above, CMP may significantly modify the characteristics of the film surface depending on the condition. On the other hand, increasing throughput is desired. The time of IPA drying may be changed according to the hydrophobic level on the surface to increase throughput. Single wafer systems are originally based on SRD. They can support both SRD and IPA drying by adding the function of Rotagoni™ drying. In this way, environmentally friendly single wafer systems with increased throughput can be provided by using an IPA drying method suitable for each type of wafer surface.

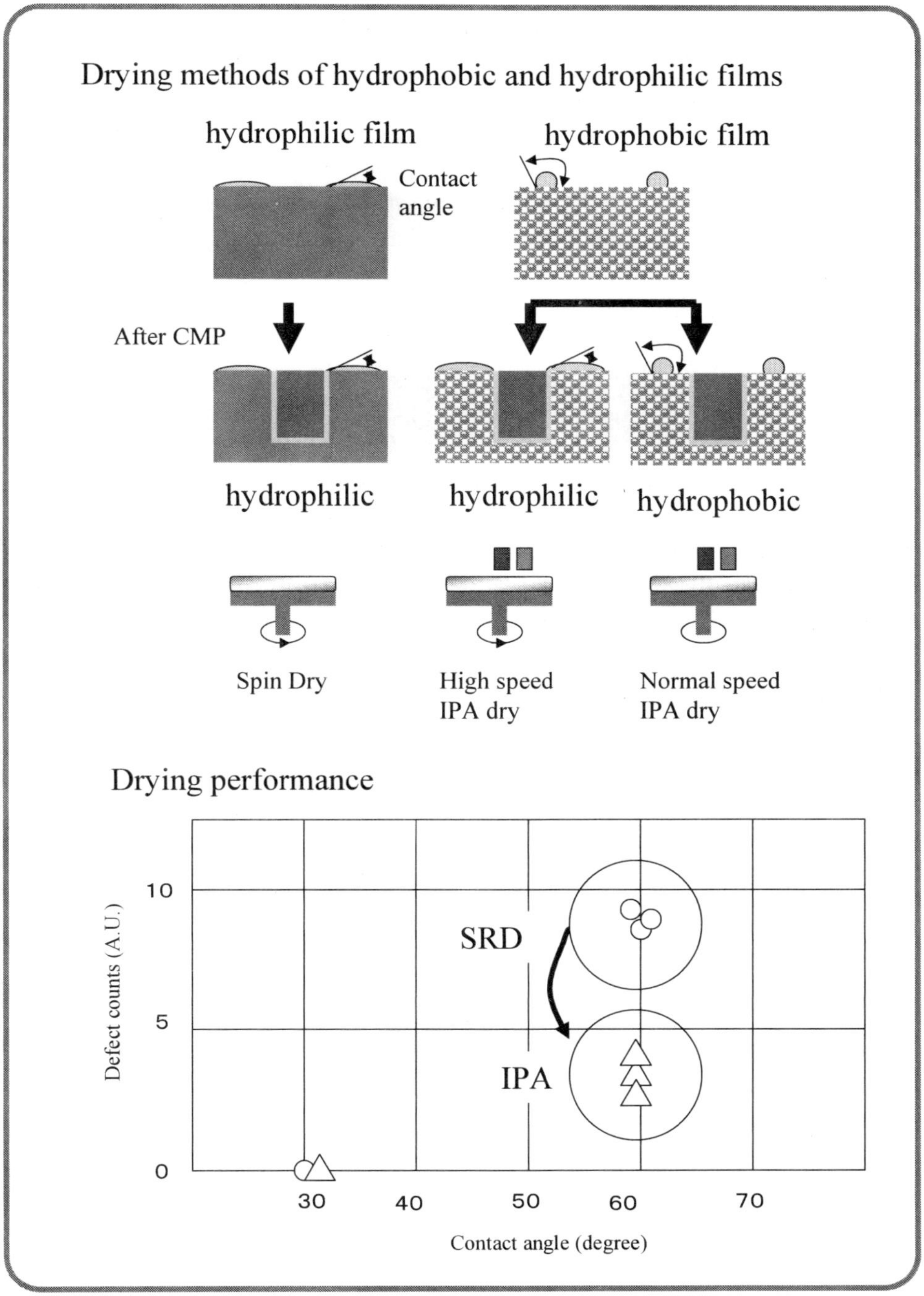
Drying methods of hydrophobic and hydrophilic films
hydrophilic film
hydrophobic film
Contact angle
After CMP
hydrophilic
hydrophilic
hydrophobic
Spin Dry
High speed IPA dry
Normal speed IPA dry
Drying performance
SRD
IPA
Defect counts (A.U.)
0
5
10
30
40
50
60
70
Contact angle (degree)

= Manabu's Law! =

The niche world of *polishing and plating* is in the midst of tremendous technological innovation. People who led the field of *polishing and plating* include Dr. Frank William Preston (1896 - 1989), known for Preston's Law, and Michael Faraday (1791 - 1867), a great natural philosopher. The ideas of Preston and Faraday have been integrated by the Japanese Einstein (that is, me) into *special and general principles of planarization technologies* described herein. *Planarization* is a permanent issue for semiconductor devices; the special and general principles of planarization technologies suggest various possible solutions for planarization in terms of deposition, removing, and combination thereof. Is it an exaggeration to say that polishing and plating follow Manabu's Law?

Dr. Moore and Dr. Dennard have led the semiconductor industry. Far back in the past, Dr. Preston and Faraday evolved polishing and plating, which were developed 3000 years ago, into science. I look forward to seeing what kinds of new technologies for polishing and plating will appear in the future and who will be the engineers (scientists) leading the semiconductor industry. Who will discover and formulate new laws? I hope that aspiring engineers and scientists do not hesitate to come on stage!

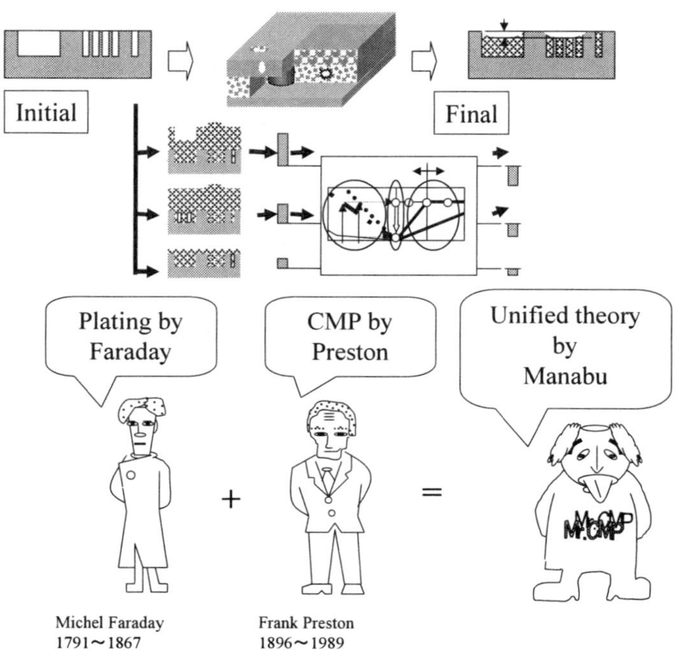

Chapter V

Wet Process Application: Interconnects

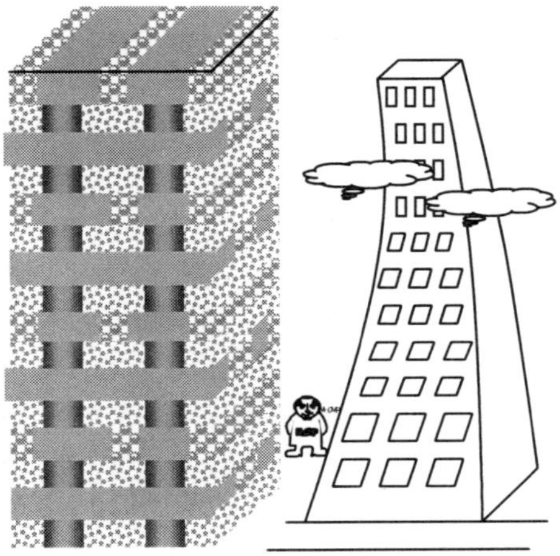

Interconnect   Skyscraper

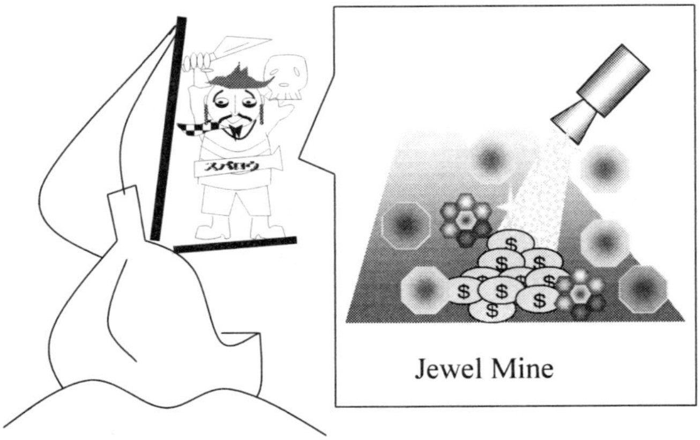

Skyscraper and multilayer interconnect problems

The figure on the next page shows multilayer interconnect problems that are often discussed in the Advanced Metallization Conference (ADMETA). Even a problem seemed to have been solved, the same one comes up again as the generation proceeds. It is like playing a whack-a-mole game. Don't be disappointed. Let's start with grasping the essence of the problems. Engineers who are engaged in the development of plating and polishing technologies are recommended to know other processes of device manufacturing.

Multilayer interconnect problems from the point of an outsider

Multilayer interconnect is used for (1) STI polishing in FEOL, (2) plating, polishing, metal cap plating, and bump plating for a damascene process, and (3) re-distribution interconnect plating in implementation. This section discusses recent problems/issues of *multilayer interconnect*. The multilayer interconnect processes include not only polishing and plating but also etching, CVD, physical vapor deposition (PVD), cleaning, annealing, ashing, and lithography. A brief explanation on the basics of multilayer interconnect problems is described below with an example of a typical damascene structure. Even engineers engaged in polishing and plating are recommended to gain the basic knowledge below. I am not an expert in devices but tried to cover every multilayer interconnect problem from the point of an outsider.

Damascene structure and issues to be solved

The multilayer interconnect processes take the following steps.

Lithography → Low-k film deposition → Cure for the low-k film → Etching → Ashing → Barrier layer deposition → Cu seed layer deposition → Cu plating → Cleaning → Annealing → Polishing → Cleaning → Deposition of DCL film or metal cap → Package

Problems that may occur in each multilayer interconnect process are discussed in the following pages. The problems include voids, rust, scratches, and low strength. There seems to be a similarity between problems of multilayer interconnects and those of high-rise apartments and skyscrapers. Let's go into details.

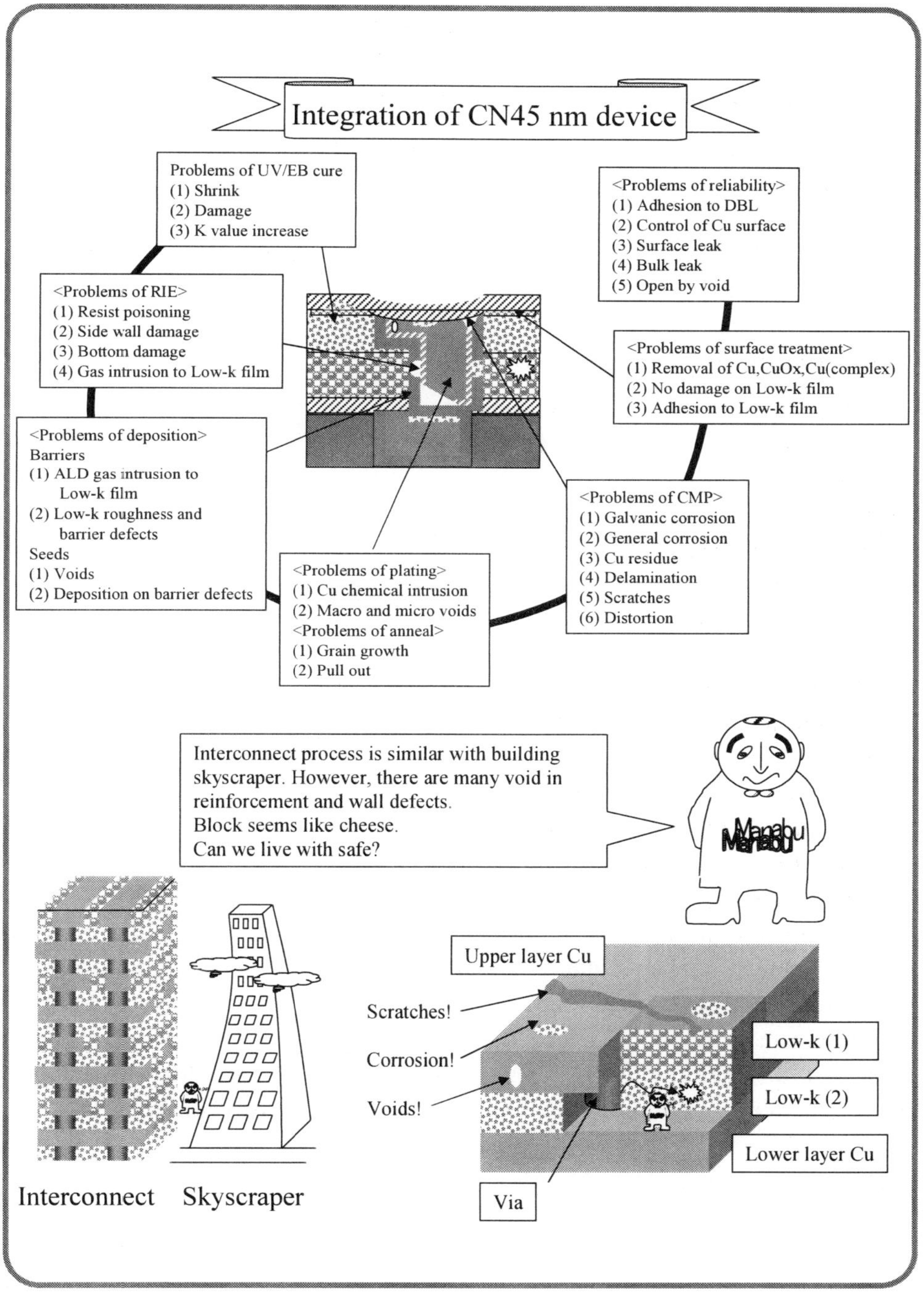
Integration of CN45 nm device
Problems of UV/EB cure
(1) Shrink
(2) Damage
(3) K value increase
<Problems of reliability>
(1) Adhesion to DBL
(2) Control of Cu surface
(3) Surface leak
(4) Bulk leak
(5) Open by void
<Problems of RIE>
(1) Resist poisoning
(2) Side wall damage
(3) Bottom damage
(4) Gas intrusion to Low-k film
<Problems of surface treatment>
(1) Removal of Cu,CuOx,Cu(complex)
(2) No damage on Low-k film
(3) Adhesion to Low-k film
<Problems of deposition>
Barriers
(1) ALD gas intrusion to Low-k film
(2) Low-k roughness and barrier defects
Seeds
(1) Voids
(2) Deposition on barrier defects
<Problems of CMP>
(1) Galvanic corrosion
(2) General corrosion
(3) Cu residue
(4) Delamination
(5) Scratches
(6) Distortion
<Problems of plating>
(1) Cu chemical intrusion
(2) Macro and micro voids
<Problems of anneal>
(1) Grain growth
(2) Pull out
Interconnect process is similar with building skyscraper. However, there are many void in reinforcement and wall defects.
Block seems like cheese.
Can we live with safe?
Manabu
Upper layer Cu
Scratches!
Corrosion!
Voids!
Low-k (1)
Low-k (2)
Lower layer Cu
Via
Interconnect
Skyscraper

Low-k materials : Like cheese?

Underlying Cu interconnects for substrates

DCL, also known as Dielectric Barrier Layer (DBL), works as a dielectric film and a barrier (cap) for Cu seed layer. The layer is referred to as 'dielectric' to distinguish it from a metal cap using metal, such as Co-W. Specifications required for this layer include (1) low dielectric constant, (2) barrier property, and (3) excellent adhesion with Cu and low-k film of a substrate.

Low-k deposition

Let's study the types of low-k materials first. Since many low-k materials are available, only a summary is provided in this section.

- SiOF (k = 3.4 to 3.8) is obtained by fluorinating SiO_2 (k = 3.9 to 4.3), the base.

- Following materials are obtained by hybridizing SiO_2 to lower the k value around 2.8 to 3.1.

HSQ $(H\text{-}SiO_3)^2$-n

MSQ $(CH_3\text{-}SiO_3)^2$-n

SiOC $(CH_3\text{-}SiO_3)^2$-n

- By forming a porous structure in the above materials to further lower the k value less than 2.5.

- The following organic low-k material has a k value around 2.6 to 3.0.

PAE (such as SiLK™, FLARE™, etc.)

- By forming a porous structure in the above material to further lower the k value less than 2.5.

Based on the above basic knowledge, let's go into details in the following pages. Film formed by either CVD or SOG (applying SOG solution) has a k value around 2.8 to 3.0 when they are dense; the k value is lowered to less than 2.5 by forming a porous structure. It is enough if you have an idea that pores are formed in a larger structure by attaching H, CH_3, or C_3H_7 which have higher molecular weight instead of O of SiO_2, and then the k value is lowered.

By the way, low-k film is like cheese, isn't it?

Damascene structure

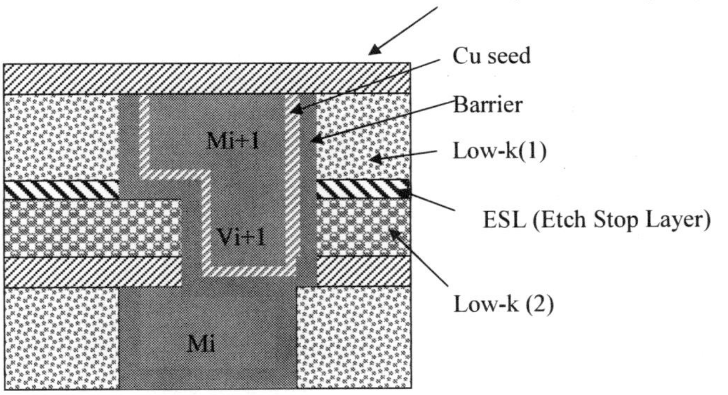

There are many problems;
(1) Low- k deposition and cure
(2) Etching and ashing
(3) Barrier and seed deposition
(4) Cu plating
(5) CMP
(6) Cleaning
(7) Cap deposition
(8) Surface treatment
(9) Packaging

Low-k film seems like cheese.

It becomes small.
Who eats my cheese?

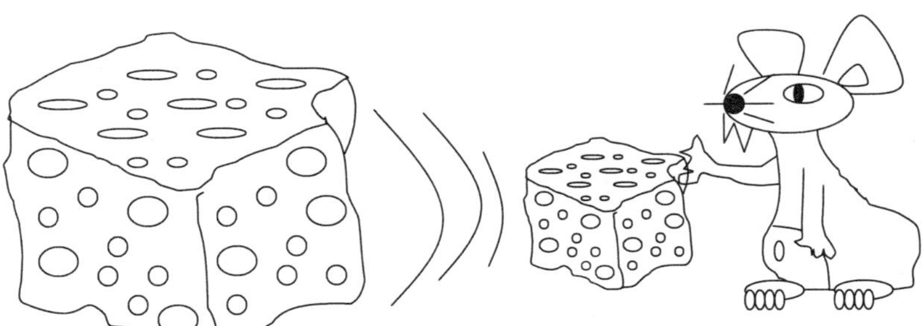

Heat cheese : Cure technology

Low-k film: Issues to be solved

Low-k film is weak. What does weak mean? The properties of low-k film are generally expressed by the following six parameters.

1) Dielectric constant: k

2) Young's modulus (modulus of elasticity): E (GPa) representing strength.

Note) 10 GPa is desirable.

3) Hardness: H (GPa)

Note) Generally, about 20 % of the value of E. For example, H will be 2 GPa when E is 10 GPa.

4) Coefficient of thermal expansion (CTE): 17 ppm/°C (CTE of Cu) is desirable.

5) Adhesion: J/m^2 Generally, about 5 J/m^2. 10 J/m^2 or more is desirable.

6) Internal stress: Low-k film generally shows compressive stress while Cu film shows tensile stress.

The problem is that lowering the k value also decreases E. In particular, porous low-k film is mechanically weak by nature, because its k value is lowered by increasing the porosity. Cure technology is used to improve strength by locally increasing the porosity at risk of increasing the k value as described below.

Cure

UV (ultraviolet)/EB (electron beam) cure processes are used to improve strength. For example, a low-k material having Young's modulus of 5 GPa can be improved to have 10 GPa by using a cure process. 10 GPa is desirable because Young's modulus of 8 GPa or higher is required for polishing. However, there are side effects. As described above, a cure process causes film shrinkage and increases the density and results in increasing the k value. The process can also damage the surface. When using a cure process to improve strength, it is essential to determine an appropriate k value while avoiding surface damage.

Film shrinkage is critical. Cure processes typically cause film shrinkage of about 10%. Film shrinkage increases stress and density and results in an increase in the k value. By balancing between strength improvement and an appropriate k value, cure processes need to be carried out effectively.

This cheese seems to shrink and get harder after heated.

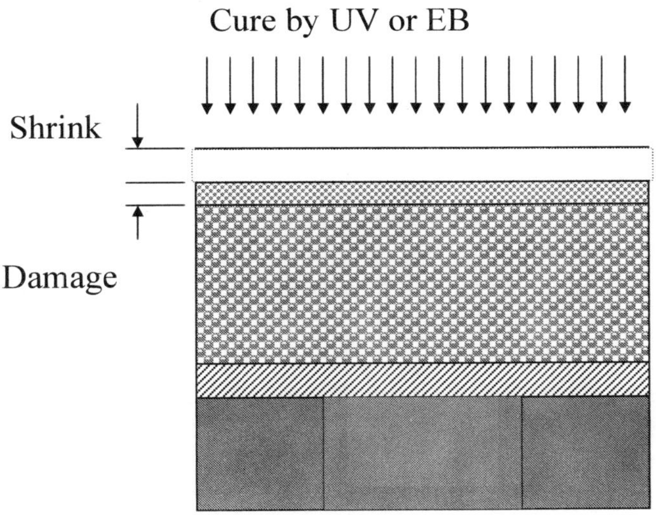

Purpose is to increase strength.
Problems are;
(1) Shrink
(2) Damage
(3) K value increase

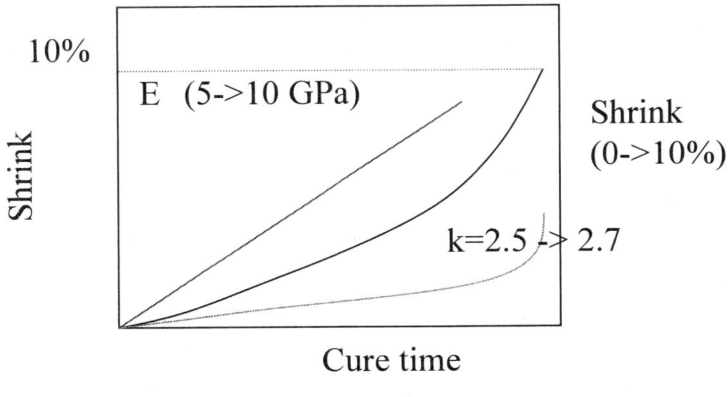

Dig, dig, and dig: Dry etching

Low-k dry etching and DCL bottom damage

Various issues arise in dry etching Major issues are as follows.

(1) Poisoning cause by nitrogen (N): Known as resist poisoning. Etching film containing N forms amines (NH_3) that diffuse into the resist. This used to be discussed as T-Top formation. A chemically amplified resist is weak to N.

(2) Sidewall damage: Issues including bowing (outward), via fence (upward material projection), and degradation of barrier performance caused by non-planar wall.

(3) Gas intrusion: Low-k film damage caused by HF produced by chemical reaction of etching gas (particularly F) intruded into low-k film and water.

Next, issues of DCL bottom damage are as follows.

(1) Cu sputtering on the sidewalls

(2) Via bottom damage (oxidization, etc.)

(3) Etching through the via bottom (OK if intentionally done such as a punch through process)

(4) Low-k film damage

Issues of barrier and seed deposition

Barrier issues depend on the method: PVD, CVD, and atomic layer deposition (ALD).

(1) Expectations for ALD and CVD are increasing since conformal deposition is not guaranteed by PVD.

(2) CVD is inferior in contamination control and adhesion. Expectations for ALD are higher but it still has issues of deposition rate and contamination.

(3) ALD gas is reported to intrude into low-k film and oxidizes barrier material.

(4) Rough low-k film and non-conformal barrier deposition cause defects.

Similar issues arise in seed deposition.

(1) Non-conformal seed deposition. Improvement is required.

(2) How holes and defects in barrier deposition should be handled? Can they be treated completely?

Barrier thickness is currently controlled at 5 nm to maintain minimum etch-stopper function. Two nm is expected in the future. Seed deposition is targeted at 5 nm or less. How come conformal deposition of thinner film is possible without defect? How can coating on rough substrate low-k film be done?

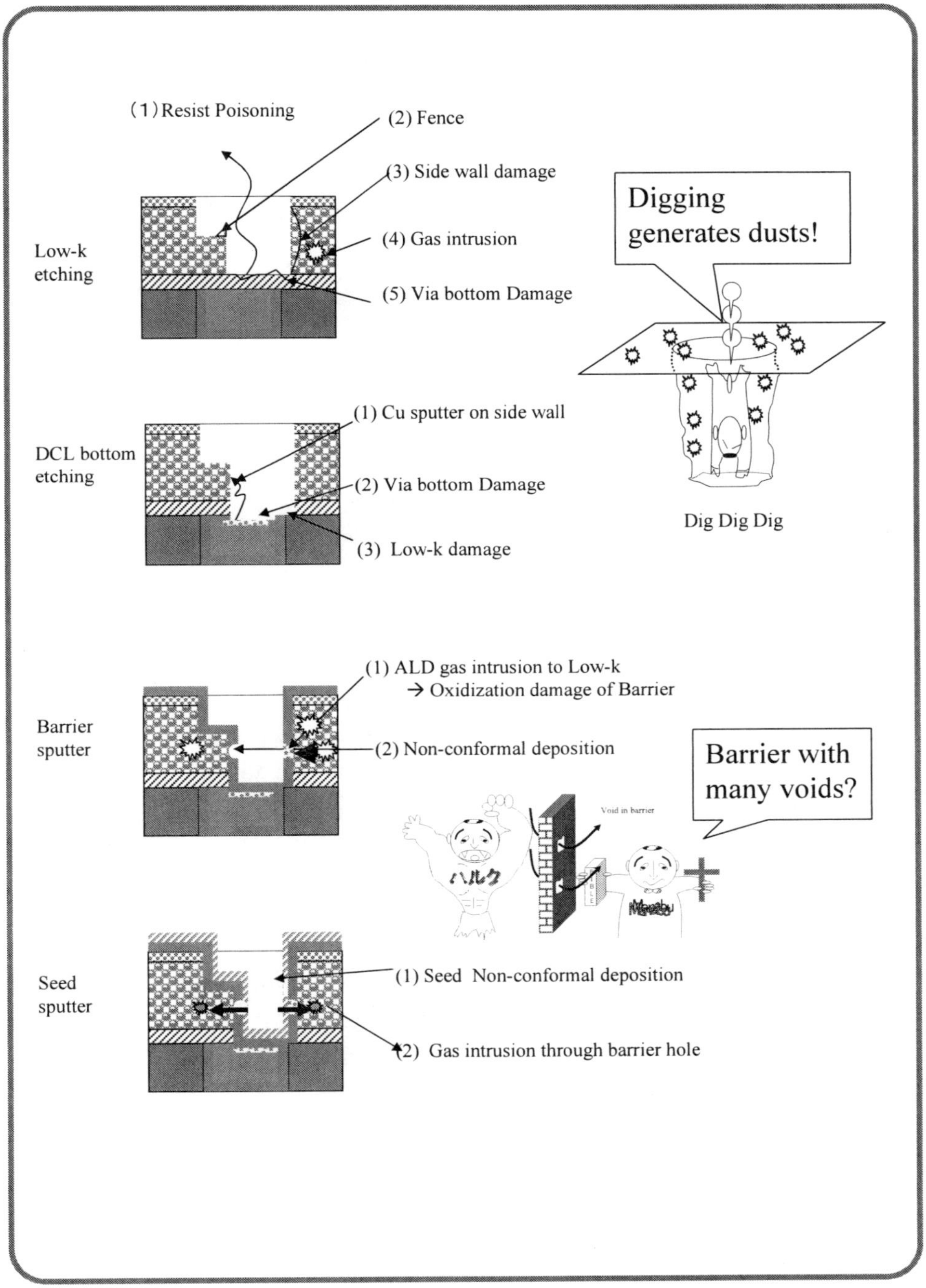
Low-k etching
(1) Resist Poisoning
(2) Fence
(3) Side wall damage
(4) Gas intrusion
(5) Via bottom Damage
Digging generates dusts!
Dig Dig Dig
DCL bottom etching
(1) Cu sputter on side wall
(2) Via bottom Damage
(3) Low-k damage
Barrier sputter
(1) ALD gas intrusion to Low-k
→ Oxidization damage of Barrier
(2) Non-conformal deposition
Barrier with many voids?
Void in barrier
ハルク
Manabu
Seed sputter
(1) Seed Non-conformal deposition
(2) Gas intrusion through barrier hole

Problems of plating and CMP

Plating: Issues to be solved

Major issues of plating are as follows.

(1) Cu solution intrusion: Cu solution intrudes through non-conformal barrier film. Cu ions were once suspected to intrude into low-k film through barrier film. No Cu ion acts like that. The presence of holes (defects) in barrier or seed film allows such intrusion. It is essential to carry out barrier and seed deposition without defect. A comprehensive solution including improvement of surface damage caused by etching in the front-end processes is essential.

(2) Voids: Macrovoids and microvoids are formed. Microvoid formation becomes a problem when EM or SM prevention is performed in the back-end processes. Macrovoids are out of the question but microvoid formation on the Cu seed surface is also a problem. Even smaller nanovoid formation on the grain boundary is more serious. Nanovoids are normally not prominent but suddenly gather into larger voids when DCL is formed and thermal cycle is given. Such voids are called stress induced voids (SIV). Cu has tensile stress during deposition but the stress changes to compressive when heated. The CTE differential between this thermal cycle and DCL decreases or increases the stress of Cu. Nanovoids start gathering according to the stress level.

(3) Purity of plating solution and EM/SM resistance: Another problem of plating to be solved. Plating solution containing more impurities is reported to enhance EM resistance but it is not a drastic solution. A drastic solution is required to eliminate EM, SM, and voids while using plating solution containing fewer impurities.

Annealing: Issues to be solved

(1) Know-how to control grain size growth in annealing is required. Larger grain sizes are better to consider a scattering problem.

(2) Via pull out: Poor annealing used to cause pulling out of the via bottom. Though this problem has already been solved, poor annealing may potentially increase stress.

CMP: Issues to be solved

Major problems of polishing are as follows.

(1) Galvanic corrosion (2) General corrosion (3) Cu residues

(4) Delamination (5) Scratches (6) Distortion

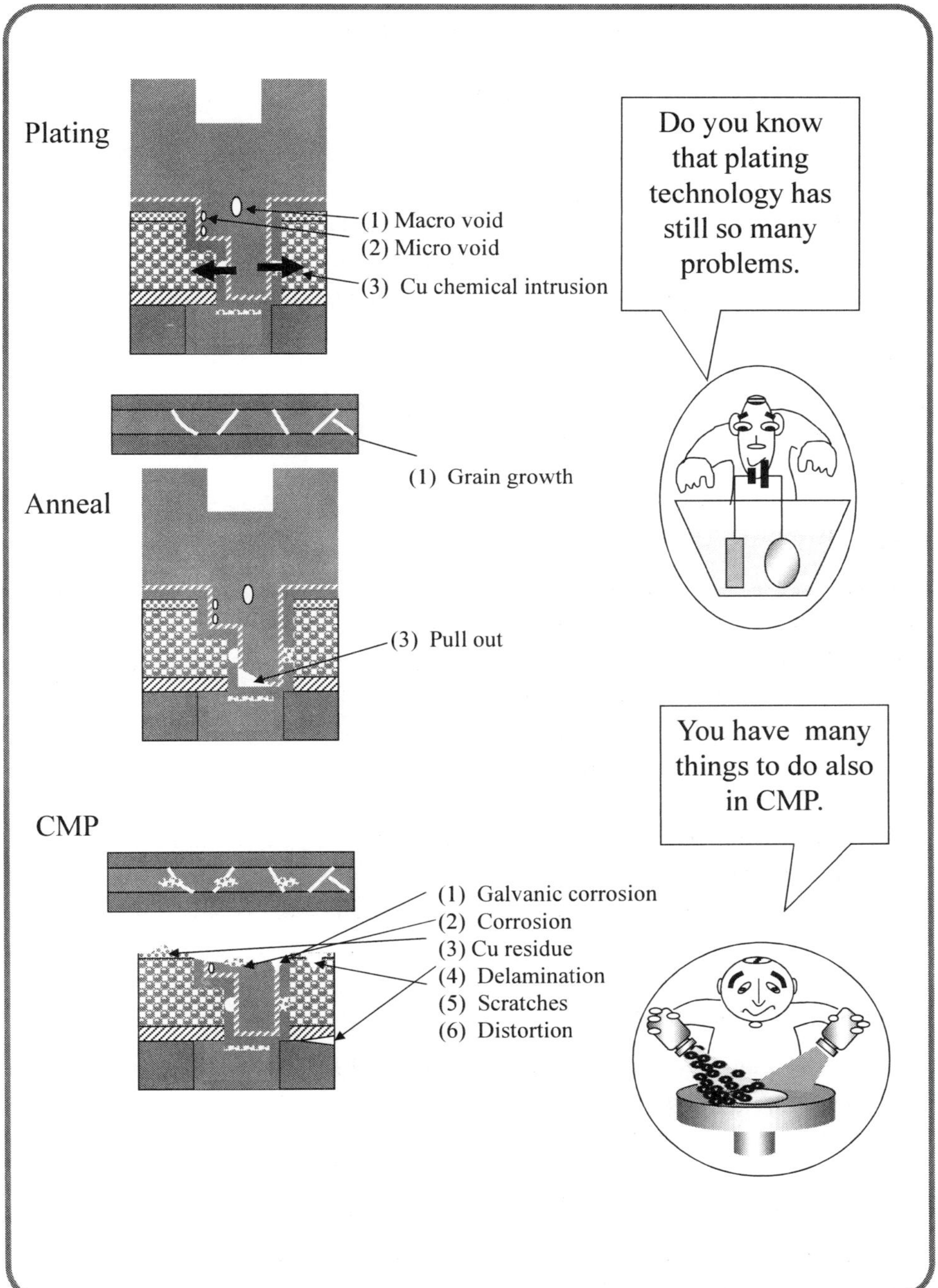
Plating
(1) Macro void
(2) Micro void
(3) Cu chemical intrusion
Do you know that plating technology has still so many problems.
Anneal
(1) Grain growth
(3) Pull out
CMP
(1) Galvanic corrosion
(2) Corrosion
(3) Cu residue
(4) Delamination
(5) Scratches
(6) Distortion
You have many things to do also in CMP.

Leak!: Reliability problem

Surface treatment: Issues to be solved

Surface treatment includes cleaning/drying after polishing and plasma treatment before DCL deposition. For cleaning/drying after polishing, light oxide film or Cu (complex) is left on the Cu surface to prevent corrosion during waiting. Cleaning and drying is performed so that Cu, CuO_X, and Cu complex are completely removed from low-k film. Watermark formation is the most serious problem in this process.

Plasma treatment before DCL deposition is discussed in various reports. Commonly used gas plasma treatment with NH_3, H_2, or He inevitably causes damage while satisfying cleaning requirement. Establishing allowable levels of damage is critical know-how.

Reliability test

To address the above problems, reliability test is conducted. As shown in the figure on the next page, five points are typically checked.

(1) Adhesion of DCL: Improvement is required (by surface plasma treatment, for example). Adhesion around 10 J/m^2 is desirable.

(2) Voids on the Cu surface: Accelerates EM/SM.

(3) Surface leak: Caused by insufficient surface cleaning or damage to low-k film.

(4) Bulk leak: Caused by insufficient barrier deposition or oxidative degradation of barrier metal.

(5) Open circuit (by voids): Many voids and surface losses are still induced by plating.

Metal cap: Issues to be solved

Metal cap plating (using such as Co-W) still has many issues to be solved. This technology is likely to be commonly used since EM resistance has already been improved.

Implementation(Package): Issues to be solved

Stress analysis in the bonding process is a major issue. Analysis of stress including (1) wire pull, (2) thermal cycle, and (3) shear stress is required.

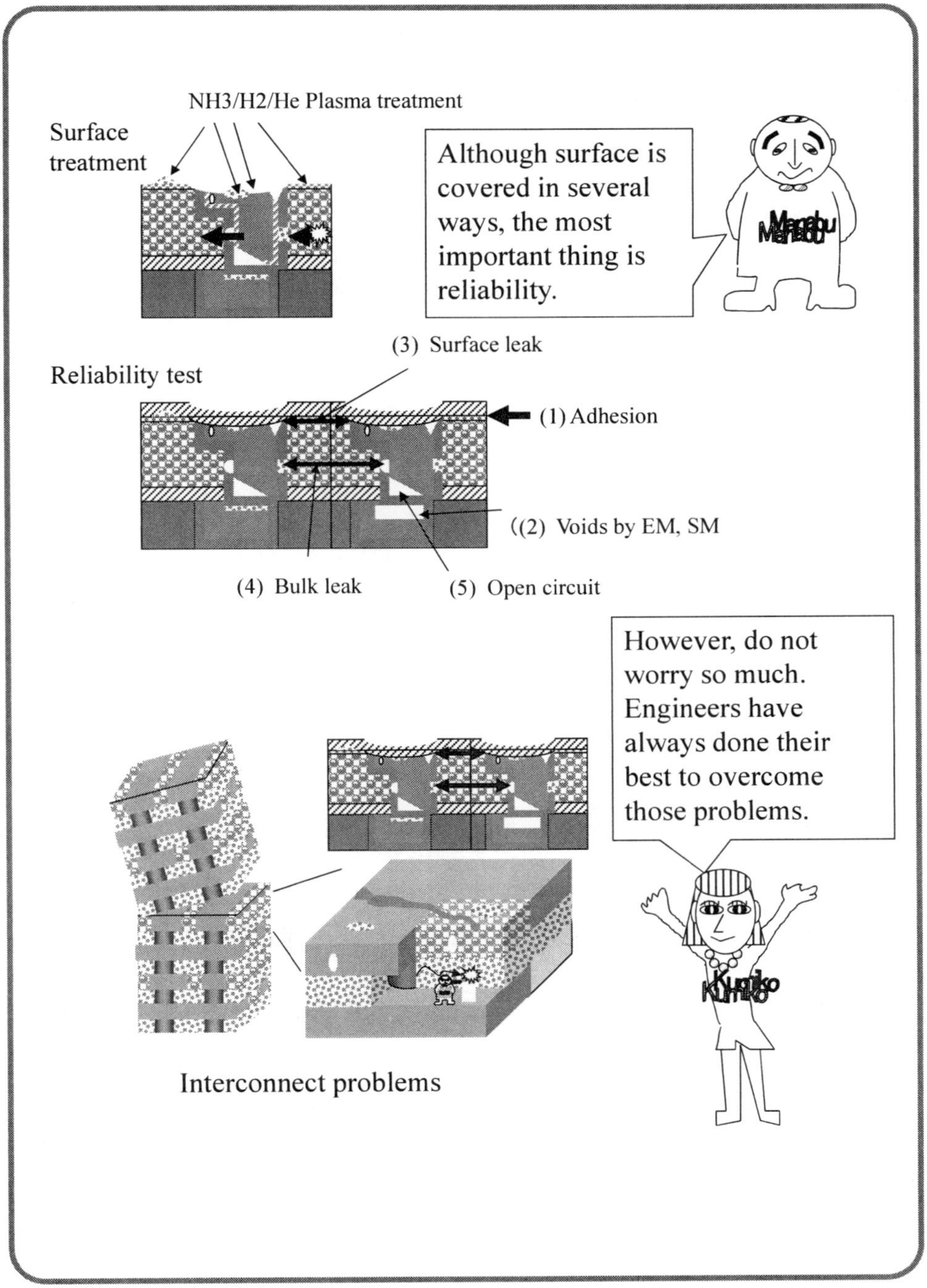
NH3/H2/He Plasma treatment
Surface treatment
Although surface is covered in several ways, the most important thing is reliability.
Manabu
Reliability test
(3) Surface leak
(1) Adhesion
((2) Voids by EM, SM
(4) Bulk leak
(5) Open circuit
However, do not worry so much. Engineers have always done their best to overcome those problems.
Kumiko
Interconnect problems

Seven issues and seven solutions

What are seven see's?

You have overviewed various problems of multilayer interconnects. Solutions to these problems are introduced below. Seven major issues extracted from users' roadmaps (including the ITRS) are shown. Seven discoveries to address seven issues are referred to as seven see's. This reference is inspired by adventures of Sinbad of the Seven Seas because some solutions involve adventurous technologies.

From requirements of roadmaps

Seven see's and promising solutions satisfying requirements of users' roadmaps are shown below.

1st see: Direct CMP of low-k layer -->

Direct polishing of porous low-k film (k = 2.5).

2nd see: Low downforce planarization

How low a downforce should be (no conclusion being reached for several years).

3rd see: Ru barrier application

Application of Ru barrier/seed, which was decided not to use for the 45 nm generation.

4th see: Watermark free

Cleaning/drying of hydrophobic film, of which k is 2.5 (in relation to direct CMP as in the 1st see).

5th see: Control of particles as small as half the size of CD

How small defects should be controlled as CD scaling proceeds.

6th see: Co metal cap

As a measure to improve EM resistance for increasing the maximum junction current density (Jmax).

7th see: Achievement of 1.5 mm EE

The above seven issues, which have been discussed for the 65 nm generation, are still and will be important for the 45 nm and 32 nm generations. It is recommended to understand these issues comprehensively.

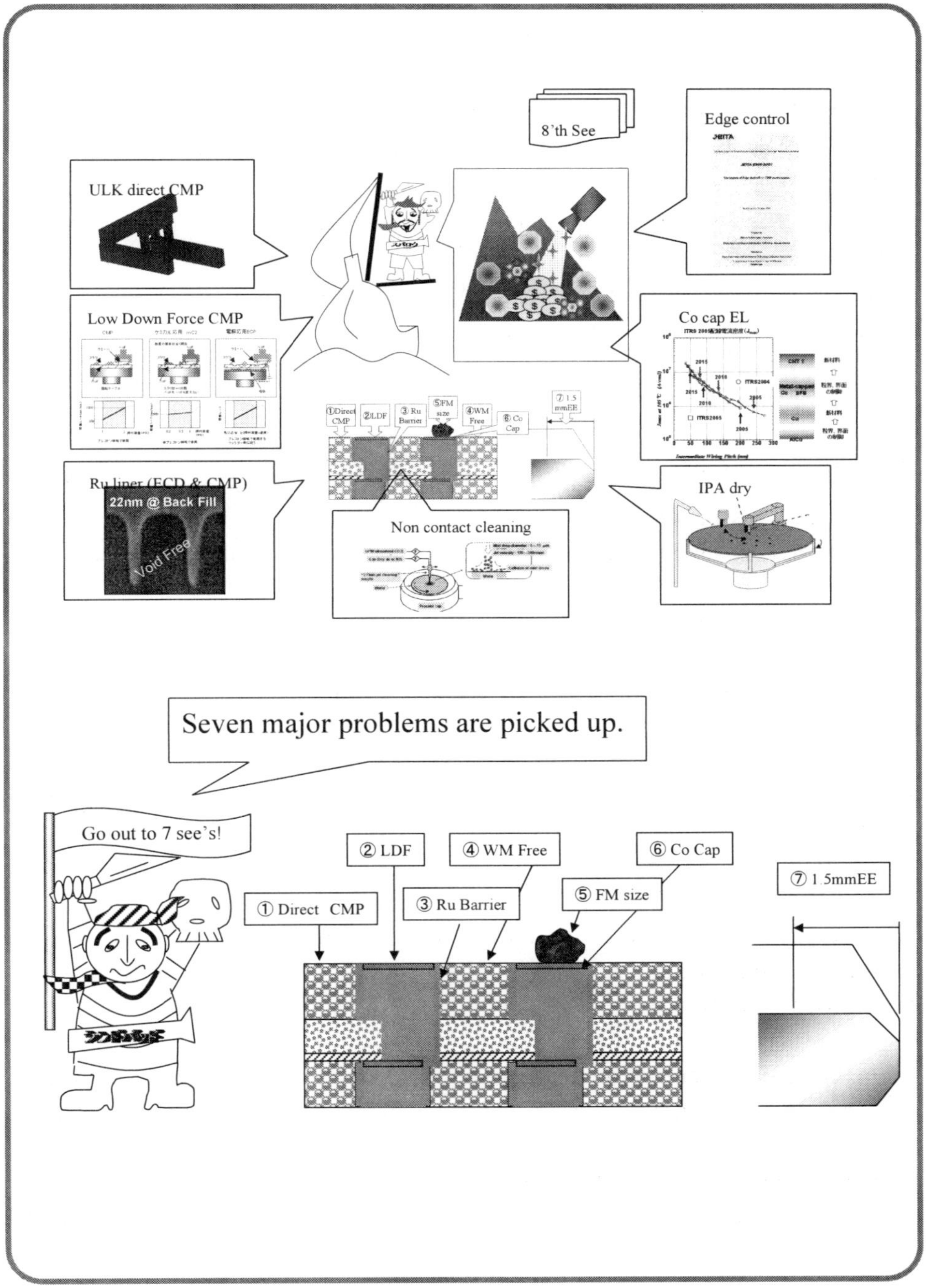
8'th See
Edge control
ULK direct CMP
Low Down Force CMP
Co cap EL
Ru liner (ECD & CMP)
22nm @ Back Fill
Void Free
IPA dry
Non contact cleaning
Seven major problems are picked up.
Go out to 7 see's!
① Direct CMP
② LDF
③ Ru Barrier
④ WM Free
⑤ FM size
⑥ Co Cap
⑦ 1.5mmEE

1st see: Direct CMP

What is the direct CMP requirement?

Three cases are currently under discussion for integration of multilayer interconnect damascene modules.

Case 1: Without a hard mask (HM); interconnects and via having the same dielectric constant.

Case 2: With a HM; interconnects and via having the same dielectric constant.

Case 3: With a HM; interconnects and via having different dielectric constants.

Case 1 is referred to as direct CMP of low-k film.

Problems and countermeasure

Let's see what the problems are.

In Case 1, damage caused by CMP as well as RIE is concerned. It is essential to develop low-k film with less damage. Damage treatment (restoration) technologies are also required.

Though damage by RIE is inevitable, Cases 2 and 3 cause less damage than Case 1 since damage by CMP can be avoided by a HM. Low-k film is intended to decrease C of RC, effective dielectric constant, keff, in integration is important. Each user will select an appropriate case comparing manufacturing technology (reliability assurance) and keff required by design. For cases using a HM, preparation of a thin HM is critical.

The middle right graph on the next page shows variation of RC when processing accuracy of HMs is varied by ± 5 nm and ± 10 nm.

Reducing RC variation is the current issue for scaling. The bottom figure shows an example of controlling variation by using an eddy current monitor in a closed-loop control system.

Direct CMP has been a never-resolved issue since organic film called SiLK™ was introduced. Organic film is effective to decrease keff but vulnerable to damage. To add or not to add a HM, that is the question for engineers. Various challenges and compromises will be coming up as improvement/development of CMP and damage treatment technologies proceeds.

Challenges for direct CMP never end.

Three cased of integration

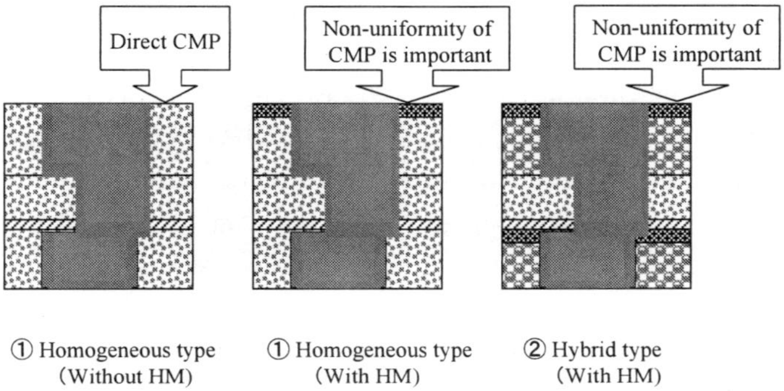

The effect on RC variation by CMP non-uniformity becomes bigger!

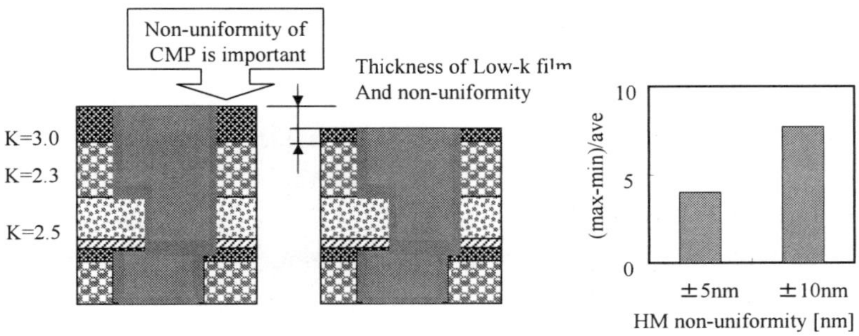

Eddy current monitor is important.

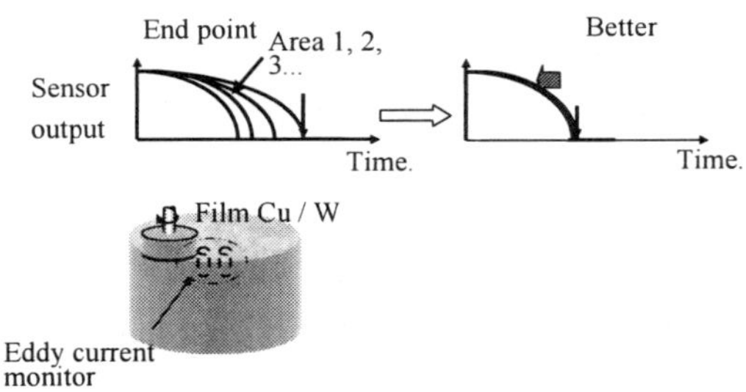

2nd see: Low downforce CMP and mC^2

What is the low downforce CMP requirement?

Strength of low-k materials is so low as to be broken by CMP; this property of low-k materials was once a driving force for low downforce CMP. Young's modulus (E) is a measure of strength. Young's modulus of 10 GPa or more was reported to be required (first mentioned in a SEMATECH report as far as I remember); the reference value of improving Young's modulus by a cure process was set to 10 GPa. Then, delamination problem was discussed. Adhesion is a measure of delamination but no clear reference value has been provided. The Gc value used in fracture mechanics or experimental data by four-point bending tests is used as measures of delamination. To address these issues, opinions to start with low downforce CMP came up. At first, a downforce as low as 0.5 psi or less was discussed and ECP was introduced. Now users select a downforce from 1 to 3 psi as CoC became a problem.

Examples of countermeasures

See the chapter "Principles of planarization technology" for details. The chapter comprehensively discusses general principles of various low downforce planarization technologies, including CMP, ECP, ECMP, and CE, as well as explanation that CMP and ECMP are same in principle, and introduction of mC^2 (a polishing method combined with etching). Taking CoC into consideration, users will select an optimal system.

ECMP versus mC^2

This section compares ECMP and mC^2 in cases of local interconnect or global interconnect.

The middle figure on the next page shows the comparison of the results of polishing by ECMP and mC^2, a type of chemically enhanced polishing (CEP). Taking CoC into consideration, users select which method to use.

After all, it is up to users which method to use. Taking CoC into consideration, select an optimal downforce from 1.5 to 3 psi.

Why LDF is necessary? So many ways.

- Planarization
- Defect
- Low-k film Distortion
  Scratches
  Delamination

are problems.

Therefore, many planarization technologies have been developed.

CMP ECP ECMP Etching Others

Unify those technologies by the general principle of planarization technologies.

The comparison of mC2 and ECMP in global interconnect

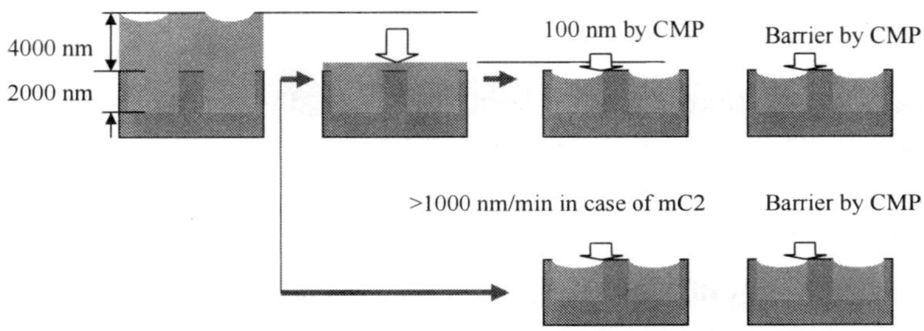

You may know its necessity by FEM stress analysis.

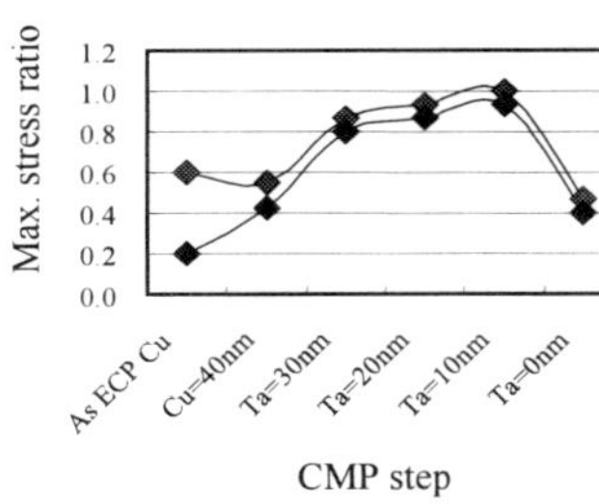

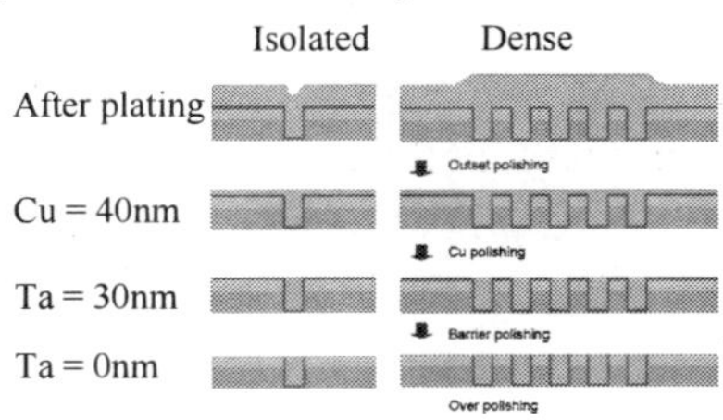

3rd see: Ru liner

What is the thinner conformal barrier requirement?

ITRS briefly introduces new materials but clearly indicates a thinner barrier as a target. A thinner barrier is essential to decrease RC. Making a barrier thinner by conventional PVD has been reported to be difficult. Instead, CVD (including ALD) has reemerged for several years since no other solution to form thinner conformal deposition was found. Recently, PVD, which forms conformal deposition by resputtering after deposition, has been introduced. In such a technological trend, Ru has emerged as a candidate for new materials.

What are issues to be solved?

A thinner Cu seed increases resistance, which increases the terminal effect [note)] and results in non-uniform deposition. Dummy resistance function or divided anodes are used to minimize the terminal effect.

In addition to a thinner Cu seed, the requirement for direct plating on Ru is already discussed.

Note) Effect of providing thicker plating on the wafer edge due to the higher resistance of the wafer center than that of the wafer edge.

Example of countermeasures

The bottom figure on the next page shows an example of the plating method that uses a dummy material on Ru. Using Ru is considered to be one of good solutions for thinner film and reliability improvement of devices.

Concerning CMP for wafers using Ru liner, CMP slurry dedicated for Ru has not really been developed yet. Once practical use of Ru liner is started, it will accelerate the development of Ru-dedicated slurry.

Filling performance will be more severe in accordance with scaling.

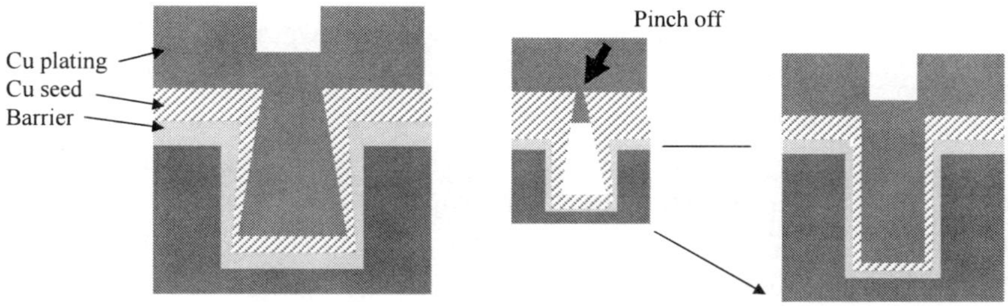

In case of bigger scaling,
① No pinch off even if PVD seed is thick.
② No terminal effect in plating.

In case of smaller scaling such as beyond 45 nm,
① Pinch off if PVD seed is thick.
② Worse filling (voids)
Therefore, thin and conformal deposition technologies for barrier and sees are required.

Example to use Ru

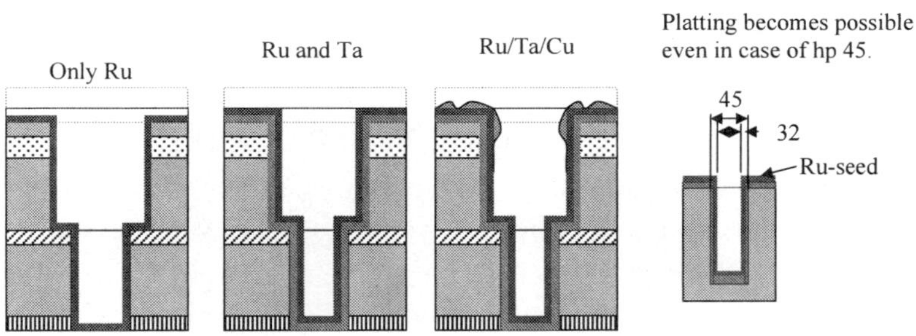

Example of plating on Ru by dummy resistance plating

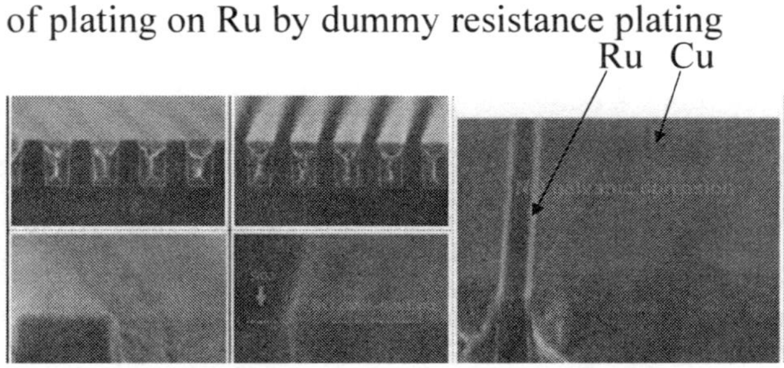

4th see: Watermark free

What is the watermark free requirement?

Watermarks must be reduced to zero. No watermark should exist. Watermark is so called *liquid glass*, expressed as SiHxOy. With the presence of water on the Si surface, Si and oxygen in the atmosphere dissolved into water to produce SiHxOy. By drying water, SiHxOy condenses into a navel-like shape. This is called a watermark. Hydrophilic film (such as TEOS) leaves no watermark. Watermarks are likely to be formed on hydrophobic film (such as ULK). IPA drying is an effective method to prevent watermarks. IPA displaces water and inhibits liquid glass formation. The top figure on the next page shows watermark formation on TEOS, dense ULK, and porous ULK, which have different hydrophobic properties. The contact angle of a droplet depends on the hydrophobic property of each film; watermark formation after drying increases proportionally to the hydrophobic property.

Example of countermeasures

An example of Rotagoni™ drying is described below.

Rotagoni™ drying was developed by IMEC. The technology applying Marangoni effect to single wafer systems is a technology superior in throughput. A Rotagoni™ drying unit supplies IPA vapor and D.I. water on the wafer surface and dries the wafer by rotating it.

A Rotagoni™ drying unit can be used for a SRD process without using IPA; IPA effect can be adjusted depending on the hydrophobic property of film. The bottom graph shows that a Rotagoni™ drying unit dries high hydrophobic film slowly and dries low hydrophobic (close to hydrophilic) film quickly. The purpose of employing a Rotagoni™ drying unit is to increase throughput by quick drying. For example, Rotagoni™ drying enhances the hydrophilic property of Poly-Si film for a flash memory after polishing and also provides excellent performance when used for a SRD process. The development of drying technology involving less CoC and more environmental consciousness should be realized by determining optimal IPA drying time while considering the hydrophobic property of film after polishing/cleaning.

Water mark in hydrophobic and hydrophilic films:

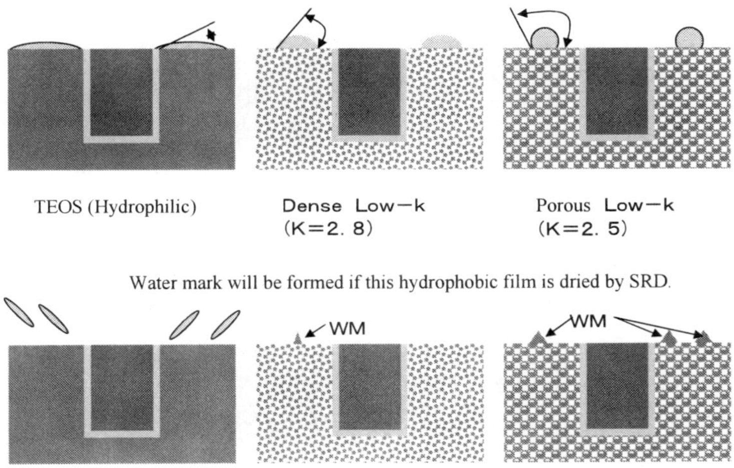

Rotagoni™ type drying technology

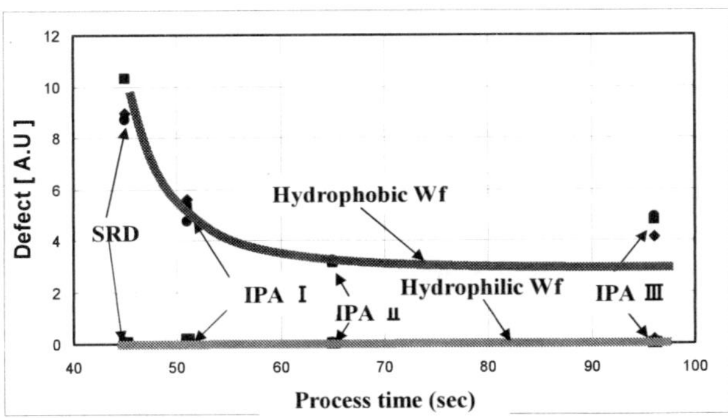

RotagoniTM type tool

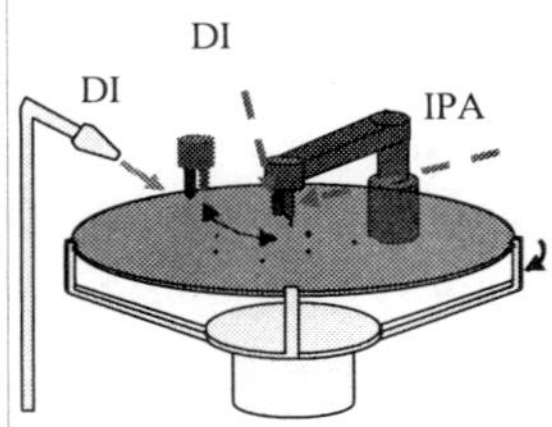

5th see: Control of particles as small as half the size of CD

What is the requirement to control particles as small as the half size of CD?

As a requirement by ITRS shows, the size of defects under control gets smaller as scaling proceeds. However, it actually depends on the detection sensitivity of measuring instruments. Currently, UV light measurement instruments can only detect approximately 100 nm (0.1 μm) particles. Some measuring instrument suppliers predict that detection of 45 nm or 32 nm particles will be possible in their roadmaps. However, based on the principle of measurement using UV light which wavelength is 193 nm, detection of 100 nm (half the UV light wavelength) particles would be minimum. Of course, detection of 45 nm and 32 nm particles realized by various technologies enhancing detection sensitivity is always welcome.

Then, how can we control *invisible particles (FM)*? Statistical method can be a solution. By applying the concept that the number of particles in a clean room is inversely proportional to the squared or cubed values of the particle size, the detectable particle size can be theoretically reduced. At any rate, for yield optimization, a simple and reliable quality control method should be considered in cooperation with users.

Examples of countermeasures

The first example is a technology to improve particle control during cleaning. In cleaning for the 45 nm generation and beyond, it is becoming more difficult to distinguish if a detected defect is a scratch or a particle. Therefore, non-contact cleaning is essential to avoid damage during the cleaning process.

The graph on the next page shows an example of particle sizes and allowable defect proportions for future generations, which were estimated on the basis of those for the 130 nm generation. This is a proposal of particle control by reducing the number of allowable defects without changing the detectable particle size even though the generation proceeds.

Inspection and processing have been playing a game of cat and mouse. In other industries, inspection completes processing. In the semiconductor industry, however, difficulty remains because it requires higher processing accuracy than inspection capability.

Scaling and FM image

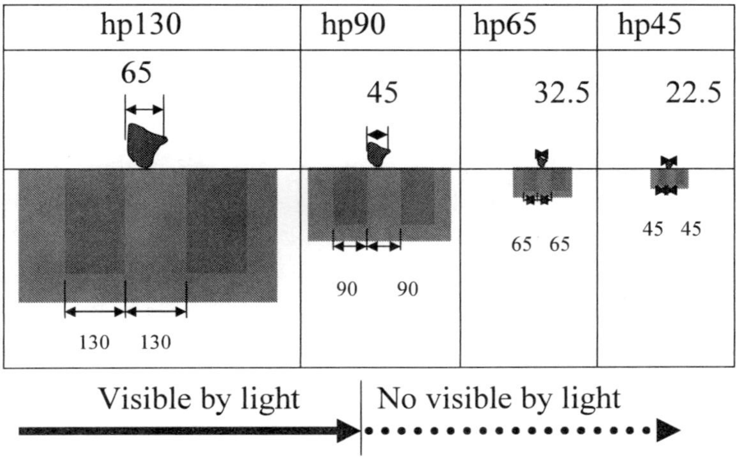

Control by statistic methods

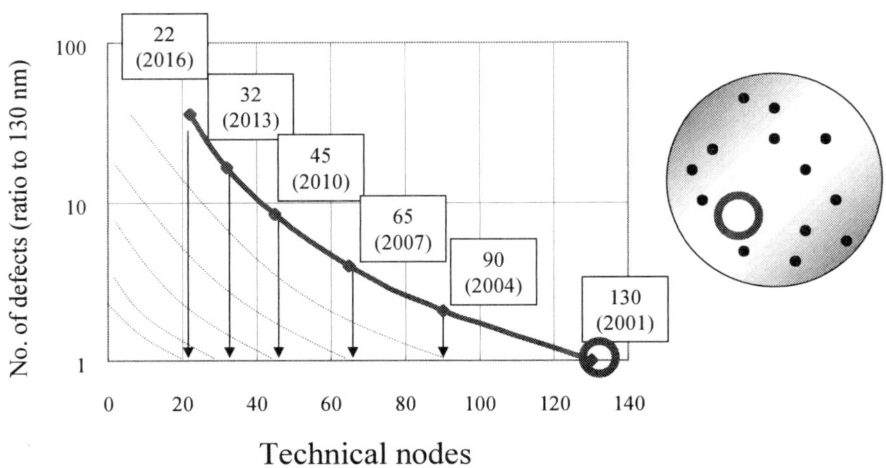

6th see: Metal cap

What is the metal cap requirement?

The top figure on the next page shows examples of integration when Co cap is used and when conventional DCL is used. Conventional DCL had a high dielectric constant (k) value; thus, reducing the k value of the dielectric layer did not result in reducing the total keff value as desired. Metal cap was originally developed with an expectation to reduce the keff value of multilayer interconnects. Now, expectation of EM resistance improvement by metal caps has become larger than that of keff reduction.

To improve EM resistance, metal caps are used with DCL. When adhesion between DCL and Cu film is low, EM resistance decreases. It is reported that the use of metal caps has improved the adhesion and made the EM resistance 100 times greater.

Example of Co cap deposition

Co cap deposition is formed by electroless plating. The types of Co cap deposition include Co-W or Co-B, depending on the plating solution. Electroless plating provides non-selective plating in nature; however, metal cap deposition needs to be formed on Cu interconnects only. To achieve such selective deposition, device manufacturers have unique techniques.

The photos show the examples of Co cap deposition with controlling the Co plating thickness to 5 to 20 nm.

Metal caps have not been widely applied. The advantage of this technique has been understood; however, some lingering hesitation has been observed in users. When the 32 nm generation comes, it will be judged whether metal caps are truly a super technology or just an excess technology.

Co cap vs DCL

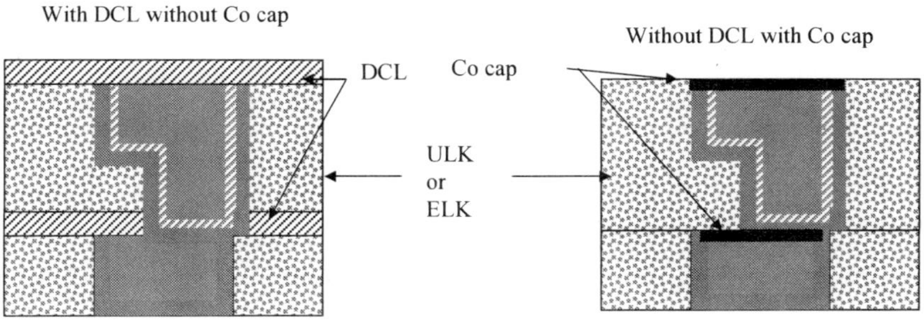

Even if k value of low-k is decreased, the effective k value is not so decreased as k value of DCL is big. If DCL is able to eliminate, k value can be much decreased.

Super technology?
Or over technology?

Example of Co cap by electro-less plating

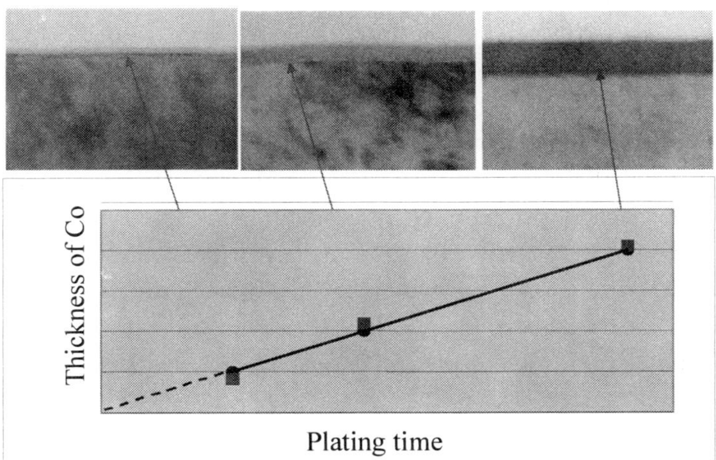

7th see: Edge exclusion

What is the 1.5 mm EE requirement?

Two mm EE of bare silicon wafers as a starting material was already required and achieved in 2005. In deposition and CMP, this requirement (2 mm EE) has not been met due to some problems, such as edge roll-off. Furthermore, the 1.5 mm EE requirement emerged in 2007.

Example of countermeasures

The 2 mm EE can be achieved with currently available ROA control techniques. In the future, to achieve 1.5 mm or less EE, careful consideration of ROA is necessary through discussions with wafer manufacturers. It is up to engineers in device manufacturers to judge whether it is absolutely necessary to manufacture devices with 1.5 mm EE from a comprehensive perspective.

PV-λ planarization range

For your reference, a handy diagram of "PV-λ planarization range," which gives a quick overview of planarization range for semiconductor devices using examples of deposition, CMP, nanotopography, and roll-off, is provided on the next page.

The x-axis indicates the topography of the film surface to be polished in wave frequency while the y-axis indicates the PV value of the film surface to be polished. The polishing line with IC1000™ pad indicates the line of 20 nm overpolishing from the peak. Thus, when conditions of the pad, slurry, and CMP are changed, this polishing line can be changed.

The diagram also shows that the device pattern is always within the range of polishing the peak and nanotopography is always within the range of not causing overpolishing. The profile of plating deposition and roll-off are in a gray area. 1.5 mm EE cannot be achieved without understanding these profiles.

Edge control

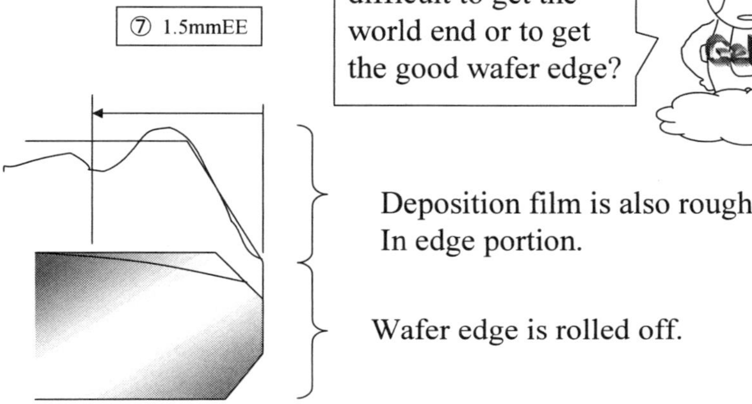

PV- λ diagram

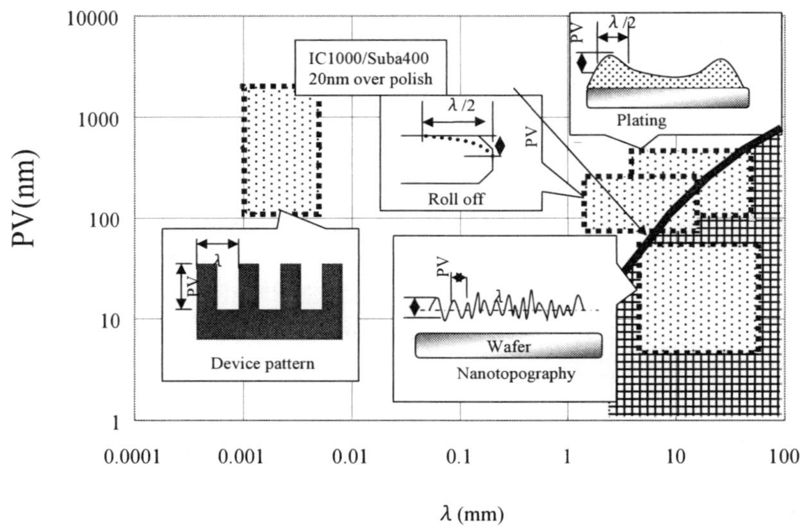

Stand-by ME

Support by ME

"Stand by Me" is one of my favorite movies, but what I mean by *Stand-by ME* is to support (stand-by) semiconductor technologies with mechanical engineering (ME). Implication of support by the author (me) is also included.

How can ME support semiconductor technologies? That is the question.

Semiconductor technologies supported by ME

Reaching technological limits in the year of 200X!

ITRS always shows directions of semiconductor technologies and also raises issues to be solved. ITRS continuously alarms technological limits of the semiconductor industry. Such concern may be similar to oil depletion, which has been long concerned but has not become reality yet. It is better for development engineers to *take technological limits as chances to spark technological revolutions*.

ME is the source of all kinds of technologies. Technological sophistication in ME would support a young industry like semiconductor in various ways. The followings are examples of useful application of ME in scaling, wafer size transition, and new materials.

- Materials mechanics and metallography for addressing the structural strength problems of new materials.
- Fluid mechanics for increasing use of wet processes (incl. polishing/plating).
- Thermal engineering for addressing thermal problems.

The semiconductor technologies are said to be a patchwork of borrowed technologies. Making best use of technologies from mature fields is critical. Engineers in other mechanical engineering fields are expected to tackle with the resolution: *how ME can support semiconductor technologies.* The followings are technological issues that *ME potentially solves*.

1) Wet processes using new materials
2) Plating interconnects with ultrahigh aspect ratio
3) Ultrahigh planarization using a damascene process (polishing)
4) Enhanced contamination control
5) Thermal removal along with increasing power consumption

They are just a part of piling issues, which can be solved with ME at nano- and micron-level.

This is literally Stand-by ME!

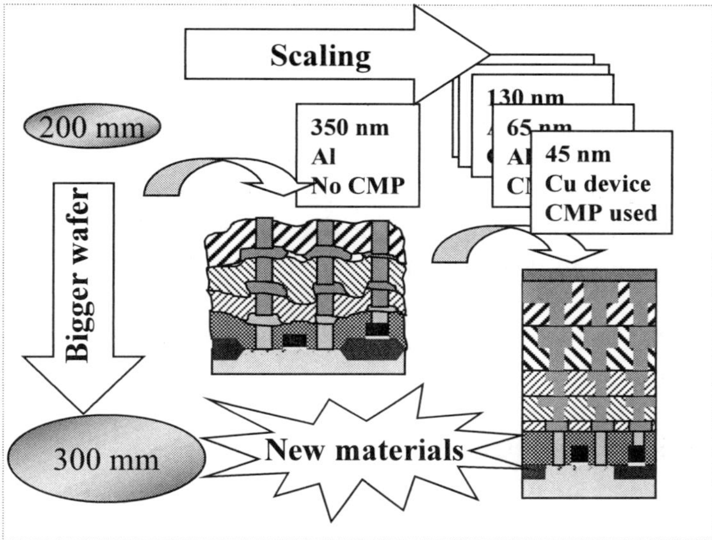

Scaling
-> Smaller FM control
FEM flow analysis
Cavitation application
-> Low-k becomes weak
FEM stress analysis

New materials
-> Corrosion countermeasure
SCC analysis

Bigger wafer
-> Precise control
Stick slip analysis
Magnetic bearing application

Semiconductor
(Baby)

Mechanical Engineering
(Adult)

FEM is all-around

FEM as a powerful analysis method

FEM is used for all categories of analysis in the field of ME. As described above with an example of analysis of damascene structures, FEM supports stress analysis at a wide range of sizes between the nano level and the kilometer level. For CMP and bonding during implementation (package) that apply load to devices, FEM can be used to determine stressed locations and the stress level at each location, helping you decide to enhance the strength of parts that may be subject to large stress.

In addition, FEM is also effective for fluid analysis and electric field analysis. It is used by engineers in all scenes.

System contamination control

Extreme cleanness is required for semiconductor manufacturing. In particular, particle control is essential. Particle control requires airflow control for which FEM is also useful. The bottom figure on the next page shows the result of analyzing airflow in a CMP system. In the CMP system, the polishing subsystem produces the largest amount of particles, but it is neighboring to the cleaning subsystem, which must be the cleanest. The air pressure and flow must be controlled to prevent the contamination of the cleaning subsystem by particles produced by the polishing subsystem.

The dry-in/dry-out concept owes its origin to FEM. Before the emergence of this concept, no one comprehended that a polishing subsystem, also called as a dust generator, and a cleaning subsystem requiring ultrahigh cleanness could be integrated. However, engineers were adventurous in any age. Using FEM, they determined that polishing and cleaning subsystems could be integrated. Then, they concentrated their efforts on achieving the goal and realized the integration of the highly polluting polishing subsystem and the ultra clean cleaning subsystem. At present, this configuration is very common.

Evolution of FEM

FEM is quite effective for stress and air flow analysis. In the following pages, the unique use of FEM for analyzing CMP/plating performance is introduced. Polishing and plating, which were known as typical 3K technologies, have come to be regarded as science with the application of FEM. This is also an achievement of engineers.

Damascene stress analysis by FEM

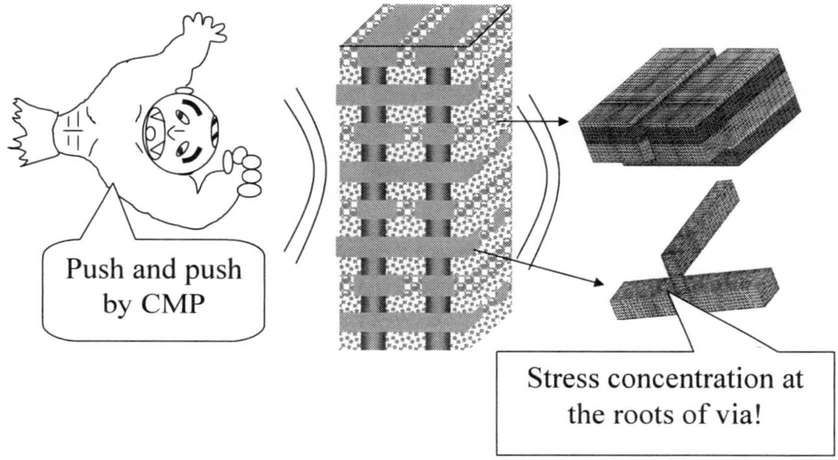

Flow analysis by FEM

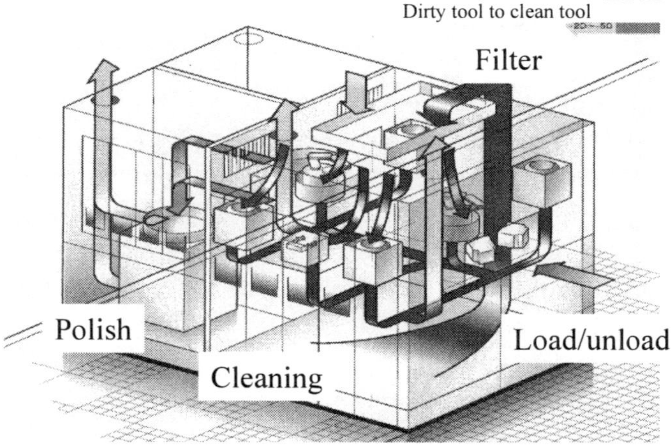

FEM for plating and CMP

Profile analysis of the polishing rate

First, the analysis of CMP for semiconductor device planarization is discussed. In semiconductor device polishing, the peculiarities of the polishing process must be grasped to provide proper profile control of the polishing rate. Analytical prediction is further developing CMP as science. Assuming that the CMP polishing rate follows Preston's equation, polishing pressures at various locations on a wafer are analyzed with FEM to provide profile analysis of the polishing rate for the entire wafer. The results obtained by FEM well agree with test results; currently, FEM is an essential technique for analyzing the CMP polishing rate.

Nanotopography analysis

Next, the analysis of wafer topography is discussed. A wafer has waviness of 10 to 50 nm in height at a wavelength of 10 to 20 mm; this topography is defined as nanotopography. In the past, the nanotopography was said to reduce STI CMP yield. STI is on such waviness. A seemingly smooth nanotopography contains a convex part, which is polished when a hard pad is used to improve CMP planarization performance. FEM analysis demonstrates that this overpolishing reduces STI CMP yield. Currently, the problem of reduced STI CMP yield can be overcome by performing CMP within the range of not causing overpolishing.

Roll-off analysis

Another current problem in wafer processing is discussed. The region of 5 to 10 mm from the wafer edge is non-uniform and has a surface level lower than the average level of the wafer surface. This region is defined as a roll-off region. The polishing profile of the roll-off region after wafer polishing is different from that of other regions, making it difficult to perform polishing control at the wafer edge. Therefore, the final polishing profile should be considered to determine the allowable level of the initial wafer roll-off.

Plating performance prediction

The electric field analysis of dummy plating has been performed with FEM for plating performance prediction.

As described above, the field of plating/polishing is supported by engineers.

Polish profile prediction by FEM

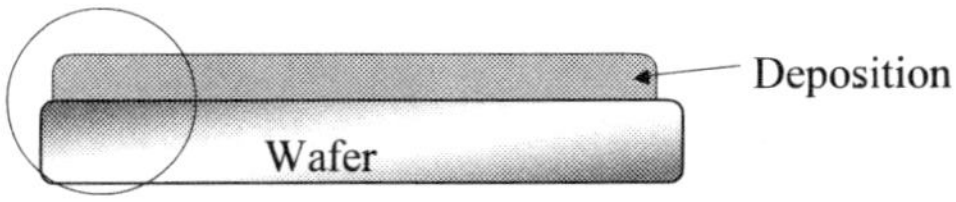

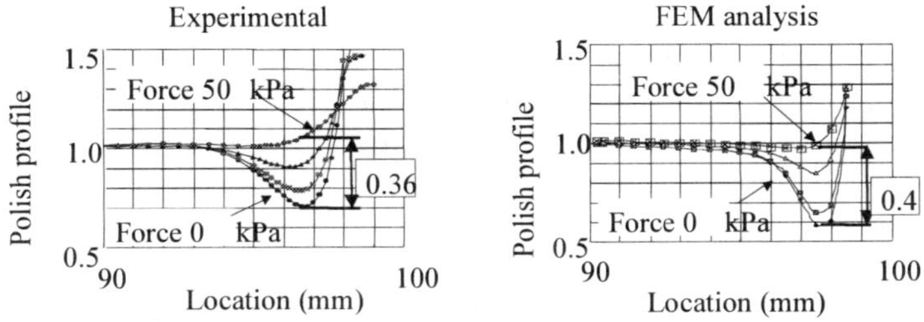

Nanotopography analysis

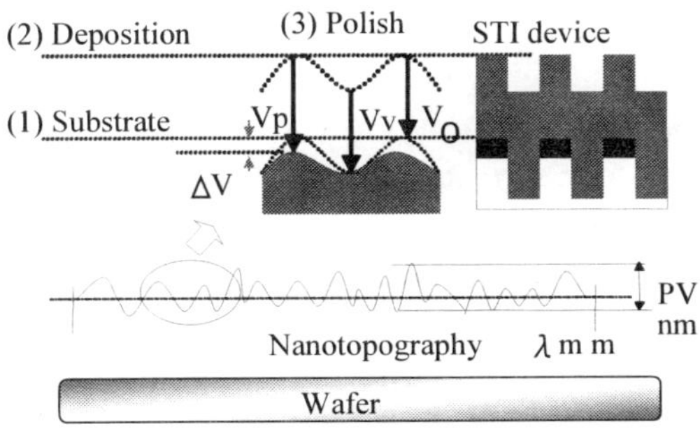

Roll off analysis

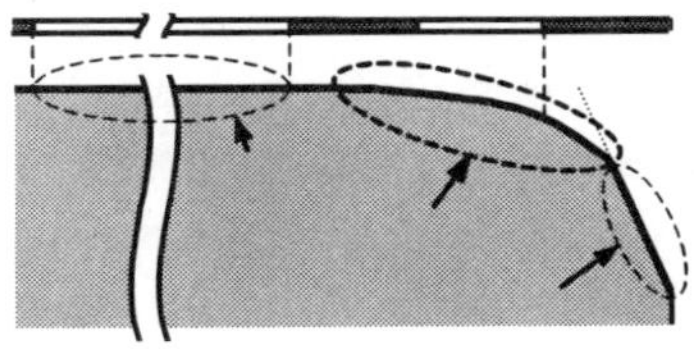

Plating performance analysis

Electric field analysis

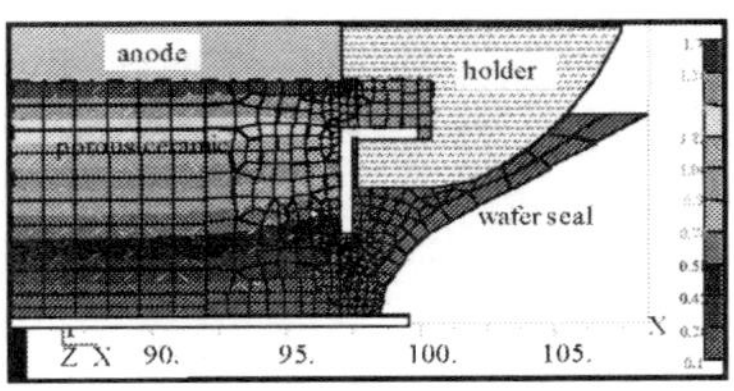

Profile analysis

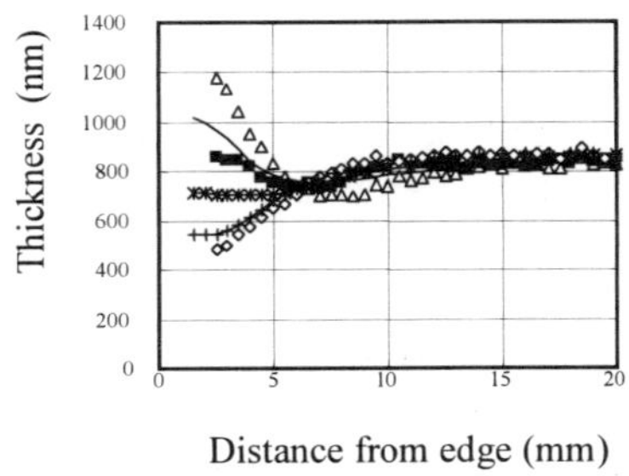

Wide application of vibration, fluid, and magnetic technologies

Stick-slip analysis (vibration mechanics)

In CMP using high load for processing (polishing), a wafer may stick and slip over the pad depending on the wafer film, pad, and slurry types, and result in vibration of the system. The top figure on the next page shows possible anti-vibration measures based on vibration analysis. Vibration analysis and anti-vibration measures can be accomplished with ME. The top graph shows a vibration generation region with the CMP rotational speed and load as parameters. Vibrations tend to occur with a lower rotational speed, higher load, and smaller slurry amount. To prevent vibrations, you may avoid operation in such a range or provide vibration prevention measures (e.g. a damper).

SCC analysis (material mechanics)

Galvanic corrosion between different materials of a damascene structure (e.g. Cu and barrier material) is well known; its acceleration by stress concentration is a little known. Analyzing stress in damascene structures and various patterns of stress applied between the Cu interconnect and barrier with FEM, a certain structure was found to promote galvanic corrosion. The photo shows actual corrosion with a FEM result. If dielectric film is of a fragile low-k material with an isolated pattern, the film sinks and the interconnect-barrier height difference promotes stress. Supposing that the stress accelerates galvanic corrosion between the Cu interconnects and the barrier, prevention measures are taken.

Cavitation jet (fluid mechanics)

This is a cavitation-based cleaning method by generating cavities with high-pressure water (about 20 atm) flown into the midst of low-pressure water (1 atm) through specially designed nozzles. Ultrasonic waves at several tens kHz generated by cavities and several tens of times wider pressure receiving area than that of conventional water jet improved cleaning performance.

Magnetic application

Magnetic bearings, as used for turbo molecular pumps and magnetic levitation systems, are applied to the CMP head and table. For the CMP head, its tilt is adjusted during polishing to control polishing performance.

Engineers make persistent efforts and fully utilize their experience for the future of semiconductor device manufacturing, which is literally Stand-by ME.

Vibration analysis

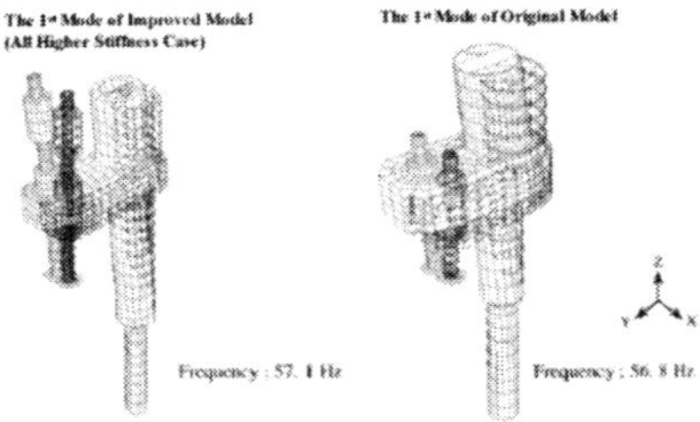

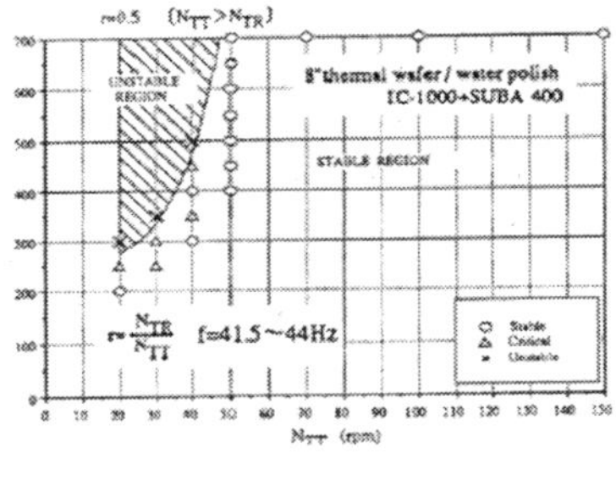

Material analysis

? Top view by SEM

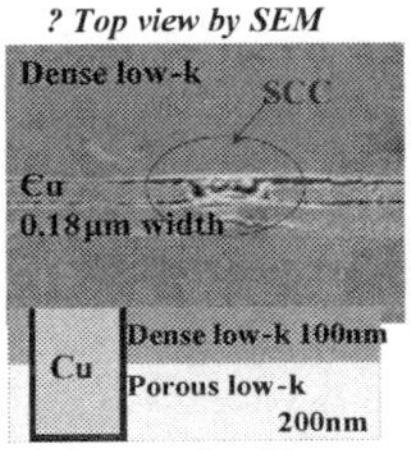

? Cross sectional image

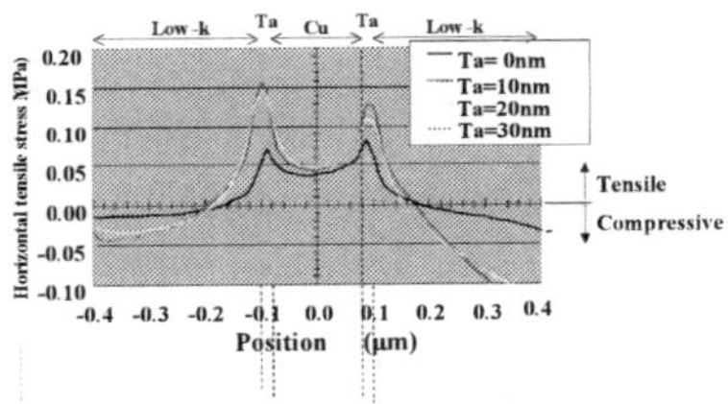

Fluid analysis

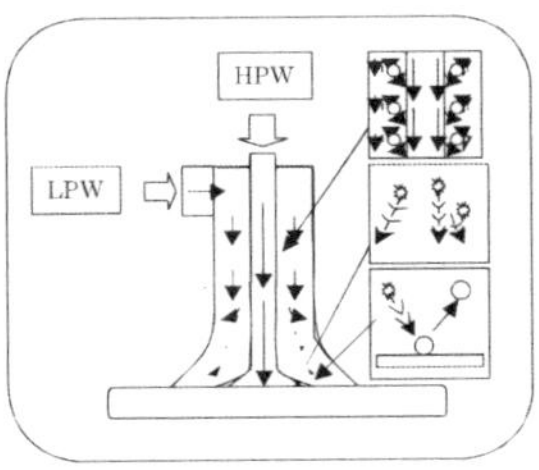

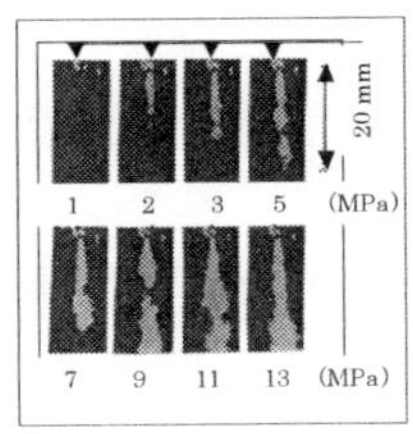

Even magnetic application

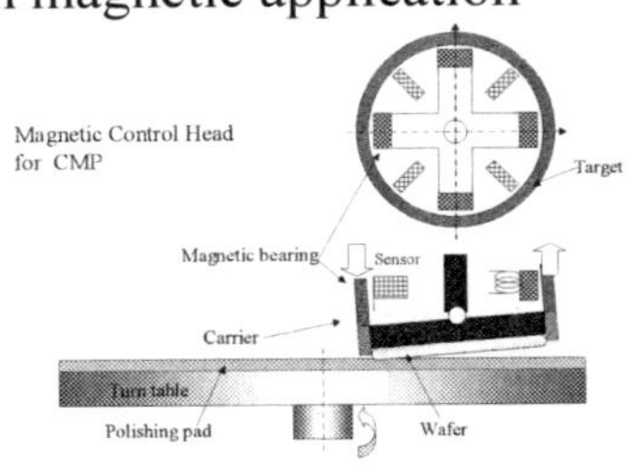

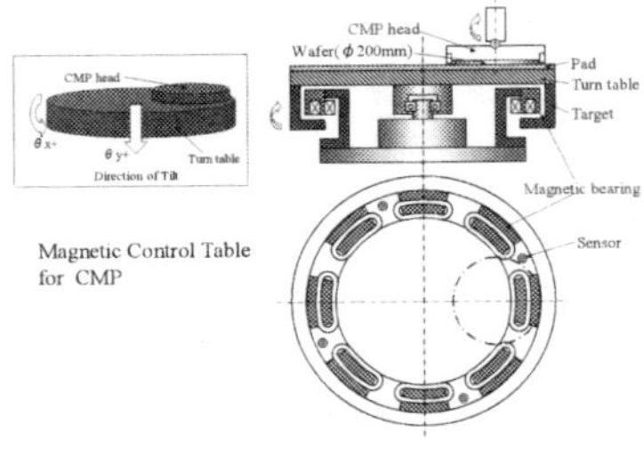

Paradigm Shift 45

No more Moore?

Let's suppose that analysts, engineers, and scientists in the semiconductor industry are having the following conversation.

The analysts say sharply, "No more Moore! (Moore's Law is not applicable any longer.) The paradigm of making profits with scaling is near its end."

The engineers retort with anger, "No! More Moore! (Moore's Law is still continuously applicable.) Feel safe that, driven by technical innovations, scaling will permanently continue."

Then, the scientists respond quietly, "More than Moore. (The solution exists where Moore's Law is overcome.) We must start to pursue emerging research devices."

In fact, intense discussions like the above are going on. About 60 years have passed since semiconductor technology was invented. The semiconductor device industry has enjoyed prosperity through scaling and wafer size transition. However, this paradigm seems to shift in the near future. I am sorry if you feel that it is far-fetched, but in the recent trend, *45* seems to be a symbolic number of breakthrough as described below. I call this hypothesis *Paradigm Shift 45*.

Paradigm Shift 45

1) 45 nm: No more Moore!
2) 45 μm: No more thinning!
3) 45 cm: No more larger wafer!

This hypothesis is first questioned by investors, who suspect whether the semiconductor industry can survive such a paradigm shift. The engineers and the scientists are obliged to answer this question.

A drastic paradigm shift is always accompanied by creation of revolution technologies. An overview of desired technologies in the future is given below in light of the symbolic number *45*.

When facing such a large-scale paradigm shift envisioned in *Paradigm Shift 45*, the engineers should reconsider the development and go back to the basics. Thus, businesses are recently placing greater expectations on universities and academic societies. The concept of *Back to the Teacher for the Future* is coming true. Efforts must be made to ensure that the semiconductor industry will not become a *lost* world after the paradigm shift.

Paradigm shifts 45

You should care about the number '45'

Paradigm shift -1 45 nm No more scaling?

Paradigm shift -1 45 um Thinner wafer for 3D package

Paradigm shift -1 45 cm Who wants bigger wafer?

P-45 discussion started from questions from investor

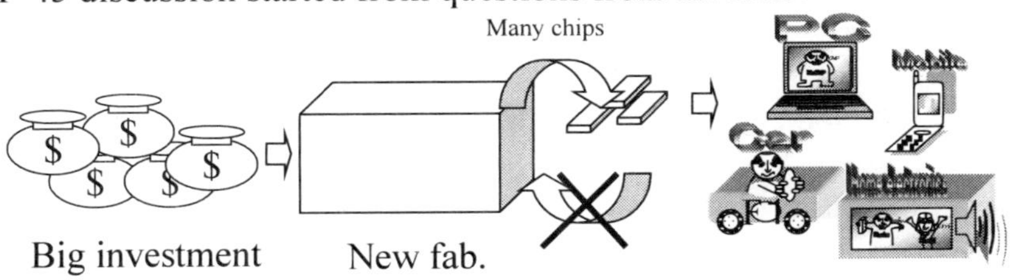

Engineers and scientists would reply to your questions.
(This is called 'No More Story')

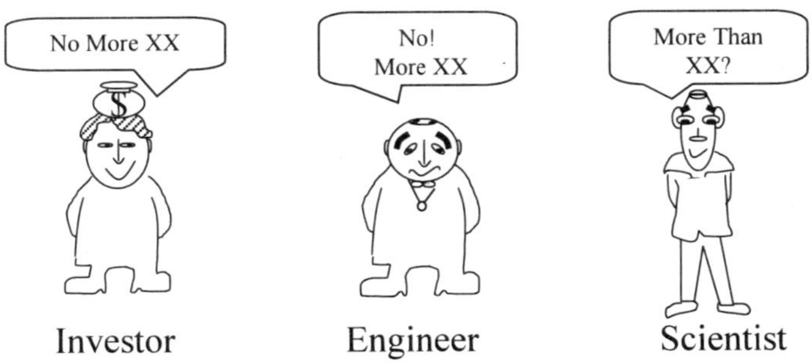

45 nm: No more Moore!

Semiconductor technologies are predicted to face engineering and physical limitations at once when scaling to the 45 nm generation is achieved. Why?

Engineering limitation (cleaning)

The cleaning technology face challenges: light cannot detect particles of 45 nm or less; the particle count exponentially increases; and the particle removal is difficult due to van der Waals adhesion. It is in an extremely difficult stage in terms of engineering. Instead of light, EB inspection has obtained attention. Visualization of particles to be cleaned accelerates our desire to removal them. A surge of innovation is coming to the cleaning and drying technologies.

Engineering limitation (processing)

The processing technology has difficulty in precise processing for the 45 nm generation. For example, the required planarization by CMP for this generation is 8 nm or less. However, the size of scratches formed in CMP is often larger than 8 nm. Can scratching be distinguished from planarization?

The defect control technology is approaching its limit; comprehensive planarization technology, including CMP, is demanded.

Physical limitation

For the generation of 45 nm or beyond, a surprising fact has come to light that the resistance of Cu interconnects is higher than that of Al. It means that the electron scattering effect of Cu interconnects provides interconnect resistance. It is literally the region where even physical constants change.

In such a region, technical innovations are seriously demanded. The application of CNT is an example of such innovation.

More Moore! (the answer from the engineers)

The engineers answer, "Scaling will continuously proceed further. Its permanence cannot be guaranteed, but technical revolution always breaks through technical limitations. Let's continue our work with confidence."

More than Moore! (the answer from the scientists)

The scientists answer, "We have already started exploring emerging research devices. They will trigger the true paradigm shift. We do not head for the end but open the door to a new paradigm."

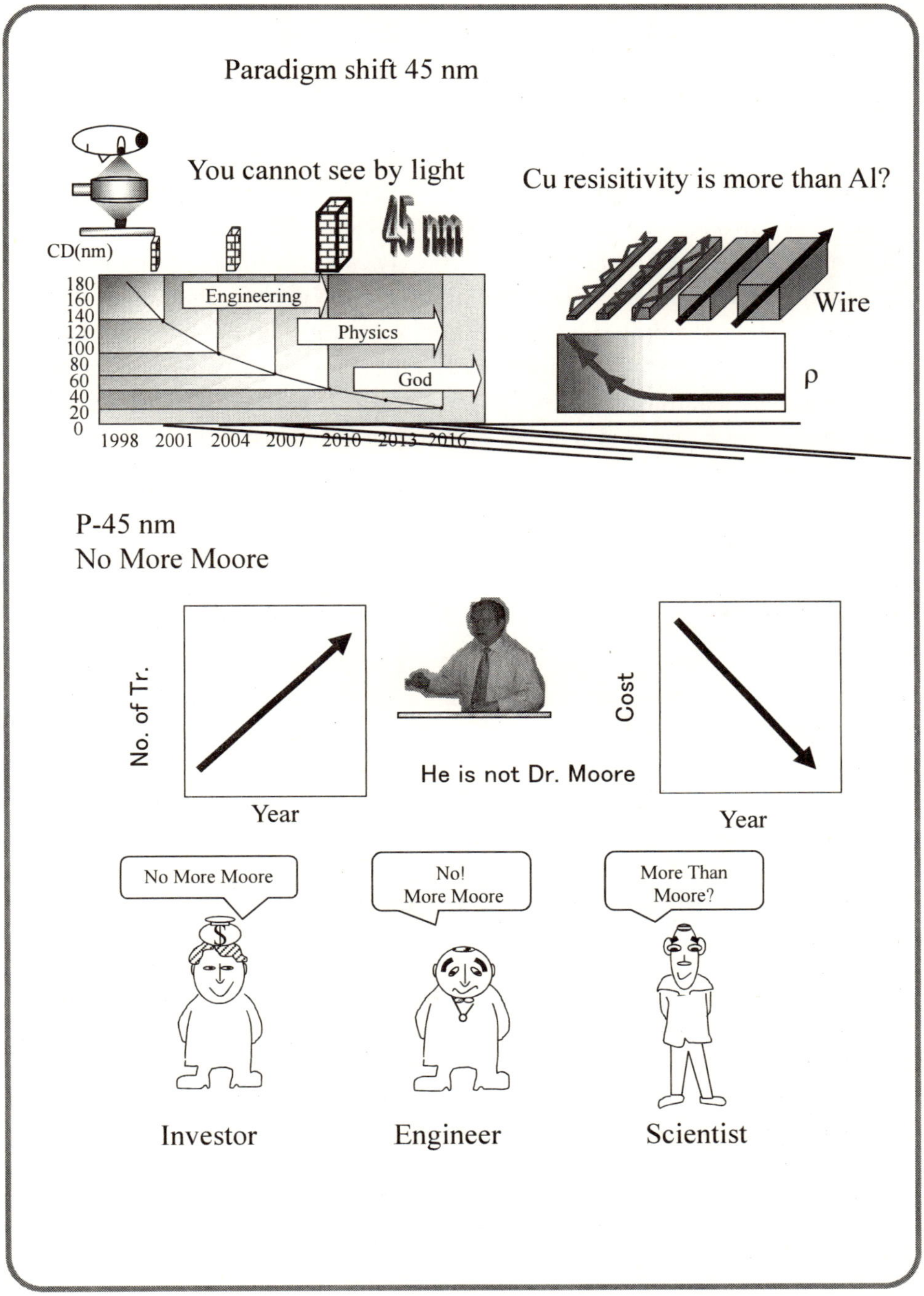
Paradigm shift 45 nm
You cannot see by light
45 nm
CD(nm)
180
160
140
120
100
80
60
40
20
0
Engineering
Physics
God
1998 2001 2004 2007 2010 2013 2016
Cu resisitivity is more than Al?
Wire
ρ
P-45 nm
No More Moore
No. of Tr.
Year
He is not Dr. Moore
Cost
Year
No More Moore
No!
More Moore
More Than
Moore?
Investor
Engineer
Scientist

45 μm: No more thinning!

Era of 3D implementation (package)

Three-dimensional implementation is being spotlighted. In the past, SoC obtained attention, and then System in Package (SiP) was proposed. The industry has currently been flooded with various terms and technologies, such as Chip on Chip (CoC) and Package on Package (PoP). Among these implementation technologies, the most important one is wafer thinning.

Wafer thinning requires substantially different wafer processing technologies depending on the wafer size. Compared to the thinning of wafers of about 100 μm or more, the handling of wafers of 45 μm or less is very difficult. Although it has already been reported that the thinning of wafers of 5 μm was achieved, there are many problems to be overcome for mass production. Thick wafers make 3D implementation troublesome in terms of processing. On the other hand, thin wafers are difficult to handle.

Introduction of 3D implementation (the answer from the engineers)

The engineers answer, "Scaling is just two-dimensional. Transition to 3D implementation is better than staying at the stage of scaling with hesitation. It has already been established for CCD devices. The 3D implementation of memory devices has also started. For such implementation of memory logic devices, many technological improvements are still required. However, it is time to proceed to 3D implementation!"

A wide range of applications of 3D implementation (the answer from the scientists)

The scientists answer, "Indeed, 3D implementation is one approach. Then, how about the possibility of four-dimensional implementation? It may suggest the direction of future development. Nonetheless, 3D implementation can be applied to transistors and memory devices. Its application to semiconductor devices is just in a nursery stage in the long history of science."

Paradigm shift 45 μm

How to make it thin in 3D PKG?

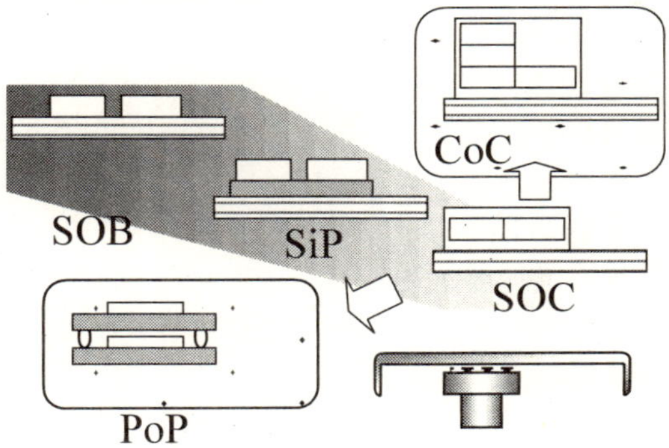

3D PKG is on going. Now is the time to start ED

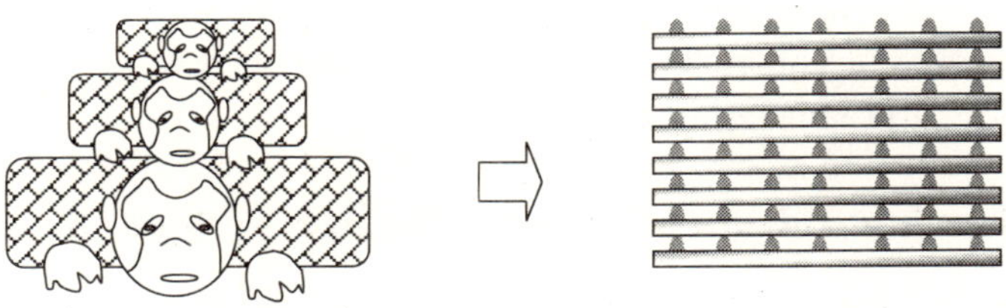

Chip on Chip   Ultra thin wafers MCM

Investor   Engineer   Scientist

45 cm: No more larger wafer!

Wafer size transition as an economic issue

Wafer size transition promotes non-uniformity in processes involving radial non-uniformity. It proceeds along with scaling. Generally, therefore, as scaling requirements become stricter, radial uniformity requirements become stricter along the transition. That doubles impact on the system development. The development of 45 cm wafers and related systems costs much. The cost rise hinders us from entering the new age. The introduction of 45 cm wafers is facing its economic limit. This situation revives the argument that took place when 30 cm wafers was introduced. The 30 cm wafers were first adopted by device manufacturers in the United States. Who takes the initiative this time?

Another case of reverse scaling?

Scaling of interconnects has proceeded along scaling of devices, but global interconnects cannot be scaled down; this problem is called reverse scaling. The wafer size transition has started with devices of older generations (with simpler specifications); however, the decrease in the wafer size for the latest devices may be a possible alternative. This is called reverse scaling of wafers (though it is a surprising compromise proposal).

45 cm wafers have been established (the answer from the engineers)

The engineers answer, "See flat panel displays. Substrates of 2 m or more has already manufactured. This case is not directly true for semiconductor devices that involve interconnect scaling, but 45 cm wafers are feasible. However, there may be economic problems."

--- (the answer from the scientists)

The scientists do not answer. "Scaling and wafer size transition should be argued by the engineers and the investors. We seek more advanced devices as emerging research devices. Preferable substrates are those that allow the fabrication of such devices. This concept reflects a four-dimensional idea."

Paradigm Shift 45 will be further argued, but will reach a conclusion someday. The engineers and the scientists are striving for creating revolution technologies to make all people in the semiconductor field happy when the conclusion is reached. The investors also expect their achievements, not just criticize them.

Semiconductor technologies are forever moving forward, I believe.

Paradigm shift 45 cm

More process variation?

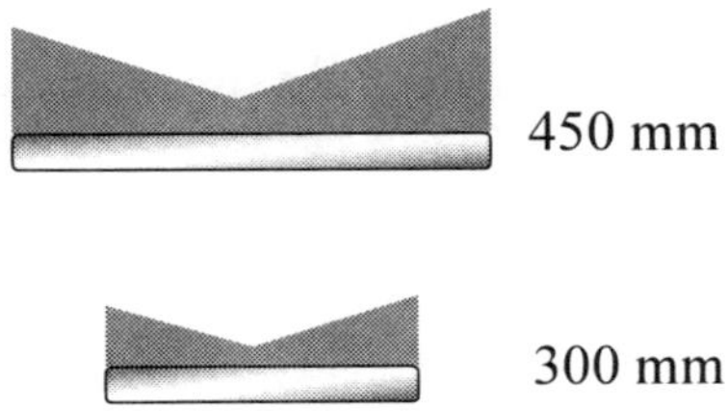

Wafer process originally has a non-uniformity in radius direction.
The bigger radius has bigger non-uniformity.
How do you overcome this contradiction?

450 mm wafer
Possible if you want!

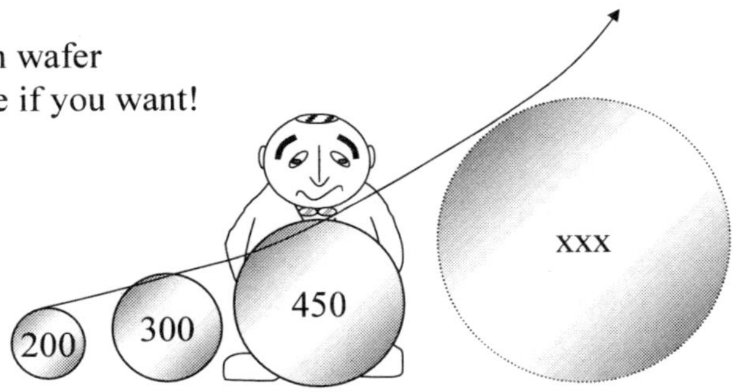

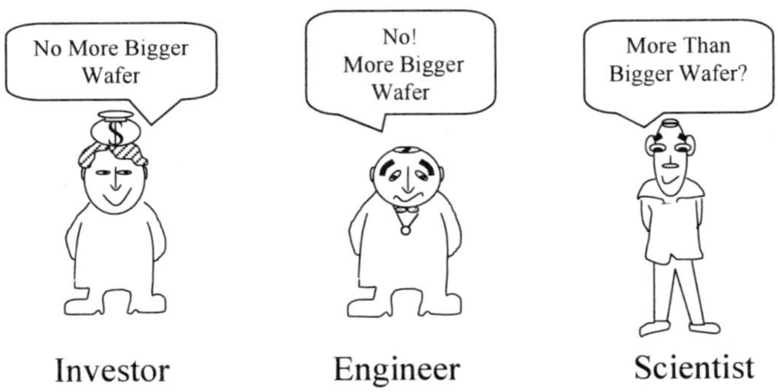

=Paradigm shift as the door to a paradise=

60 years have passed since semiconductor technology was developed. The semiconductor industry has enjoyed success through repeated scaling and wafer size transition, as well as adoption of new materials. To promote technical innovation, the ITRS has continuously raised issues related to technical barriers. As described above, Paradigm Shift 45 is becoming reality today, and the combination of all technical barriers in scaling, wafer size transition, and implementation is emerging at the same time! Investors are continuously making critical comments, while placing expectations on the semiconductor industry. Engineers are expected to create technological innovations and be proud to do so. Scientists see the paradigm shift as a good opportunity to enable the further evolution of semiconductor technology. After overcoming all difficulties, the door to a paradise may open. Please read the text above for details.

The depletion of oil has long been predicted, but has not yet become reality. The scaling of semiconductor devices has also been predicted to end, but it still continues in fact. Thanks to the never-ending efforts of engineers and scientists, semiconductor technology is forever moving forward as long as it is supported by these people.

Plating and CMP are also evolving at a rapid pace. Even these traditional technologies can be applied to the high-tech semiconductor industry. Wet process technology has been attracting attention and wonder. It is promising that amazing innovative technologies will emerge and bring breakthroughs in the future. Wet process technology is also forever moving forward.

I am engaged in the field of wet process technology, and I am also forever moving forward.

Finally, I sincerely hope that this book will provide valuable guidance and will be used by many people. Your comments on this book would be greatly appreciated. I am looking forward to having another opportunity to share fun with you.

Paradigm shifts to Paradise shift